HOME SKILLS

Plumbing

INSTALL & REPAIR YOUR OWN TOILETS, FAUCETS, SINKS, TUBS, SHOWERS, DRAINS

COOL
SPRINGS
PRESS
Home and Garden Experts™

MINNEAPOLIS, MINNESOTA

CONTENTS

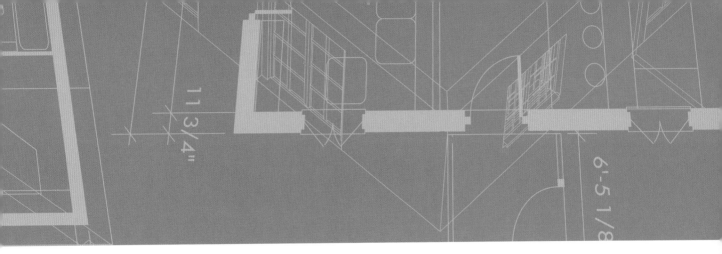

Plumbing Repairs & Projects

References

Introduction

EVERY HOMEOWNER SHOULD have a good basic understanding of the plumbing system in their home. *HomeSkills: Plumbing* guides you through the layout of the standard plumbing system, the materials used, both past and present, the tools necessary for repairs and projects, and numerous projects and repairs you can easily accomplish.

The plumbing system, like the electrical system, may seem deeply mysterious at first glance. How does this seemingly jumbled assortment of pipes (especially in an older home that may have been updated many times) make any sense. You will see that it actually does make a good deal of sense.

While not as dangerous as the electrical system, improperly installing or repairing plumbing systems can cause serious damage to your home through water leakage. It is important to follow current code and building practices to assure a dry, functional installation.

THE HOME PLUMBING SYSTEM

The Home Plumbing System

Meter

Key-operated shutoff (open)

An outdoor main shutoff may be as simple as an exposed valve that you turn by hand. Or it may be buried in a housing that is sometimes called a Buffalo box. In this example, both the meter and the main shutoff are housed in the box; in other cases, the meter is located inside the house.

BECAUSE MOST OF a plumbing system is hidden inside walls and floors, it may seem to be a complex maze of pipes and fittings. But the information in this book will help you gain a basic understanding of your system. Understanding how home plumbing works is an important first step toward doing routine maintenance and money-saving repairs.

A typical home plumbing system includes three basic parts: a water supply system, a fixture and appliance set, and a drain system. These three parts can be seen clearly in the photograph of the cut-away house on the opposite page.

Fresh water enters a home through a main supply line (1). This fresh water source is provided by either a municipal water company or a private underground well. If the source is a municipal supplier, the water passes through a meter (2) that registers the amount of water used. A family of four uses about 400 gallons of water each day.

Immediately after the main supply enters the house, a branch line splits off (3) and is joined to a water heater (4). From the water heater, a hot water line runs parallel to the cold water line to bring the water supply to fixtures and appliances throughout the house. Fixtures include sinks, bathtubs, showers, and laundry tubs. Appliances include water heaters, dishwashers, clothes washers, and water softeners.

Toilets and exterior sillcocks are examples of fixtures that require only a cold water line.

The water supply to fixtures and appliances is controlled with faucets and valves. Faucets and valves have moving parts and seals that eventually may wear out or break, but they are easily repaired or replaced.

Waste water then enters the drain system. It first must flow past a drain trap (5), a U-shaped piece of pipe that holds standing water and prevents sewer gases from entering the home. Every fixture must have a drain trap.

The drain system works entirely by gravity, allowing waste water to flow downhill through a series of large-diameter pipes. These drain pipes are attached to a system of vent pipes. Vent pipes (6) bring fresh air to the drain system, preventing suction that would slow or stop drain water from flowing freely. Vent pipes usually exit the house at a roof vent (7).

All waste water eventually reaches a main waste and vent stack (8). The main stack curves to become a sewer line (9) that exits the house near the foundation. In a municipal system, this sewer line joins a main sewer line located near the street. Where sewer service is not available, waste water empties into a septic system.

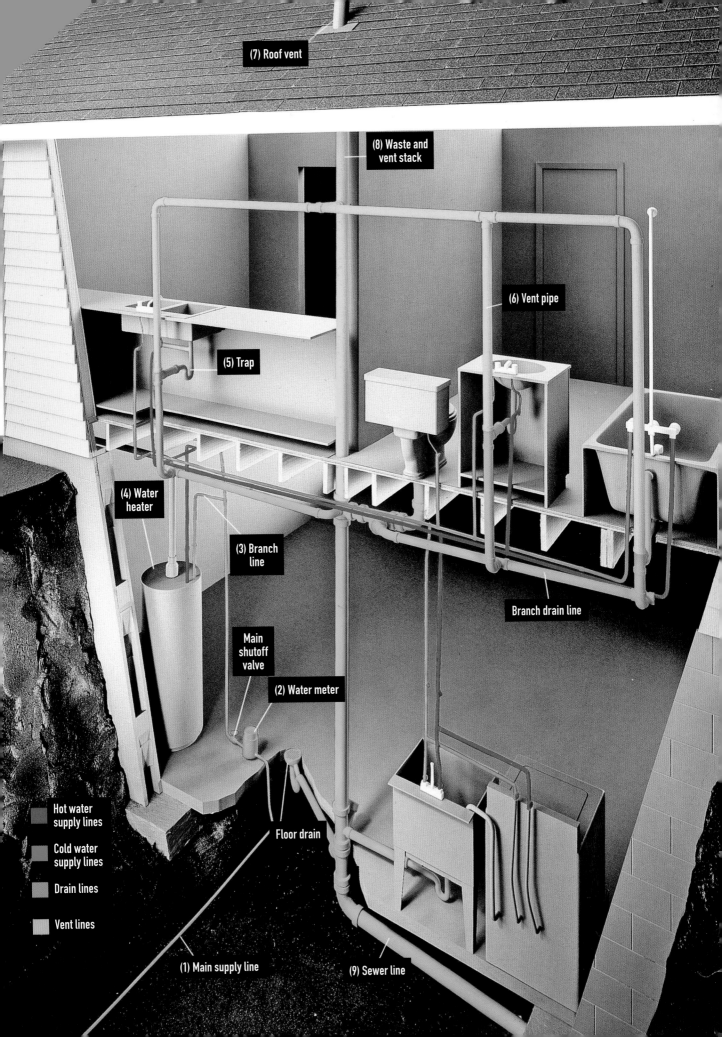

(7) Roof vent

(8) Waste and vent stack

(6) Vent pipe

(5) Trap

(4) Water heater

(3) Branch line

Main shutoff valve

(2) Water meter

Branch drain line

Hot water supply lines

Cold water supply lines

Drain lines

Vent lines

Floor drain

(1) Main supply line

(9) Sewer line

You may have an inside main shutoff, usually located near the point where the main supply pipe enters the house near the water meter. There may be valves on each side of the meter; turn off either one of them to shut off water to the house. The copper grounding wire is an important part of the electrical system and should never be removed.

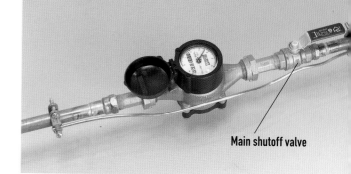

Main shutoff valve

Water Supply System

Water supply pipes carry hot and cold water throughout a house. In homes built before 1960, the original supply pipes were usually made of galvanized steel. Newer homes have supply pipes made of copper. Beginning in the 1980s, supply pipes made of rigid plastic (PVC or CPVC) became more commonplace, and the more recent plumbing innovations find PEX pipe widely used and accepted.

Water supply pipes are made to withstand the high pressures of the water supply system. They have small diameters, usually ½" to 1", and are joined with strong, watertight fittings. The hot and cold lines run in tandem to all parts of the house. Usually, the supply pipes run inside wall cavities or are strapped to the undersides of floor joists.

Hot and cold water supply pipes are connected to fixtures or appliances. Fixtures include sinks, tubs, and showers. Some fixtures, such as toilets or hose bibs, are supplied only by cold water. Appliances include dishwashers and clothes washers. A refrigerator icemaker uses only cold water. Tradition says that hot water supply pipes and faucet handles are found on the left-hand side of a fixture, with cold water on the right.

Because it is pressurized, the water supply system is occasionally prone to leaks. This is especially true of galvanized iron pipe, which has limited resistance to corrosion.

For some houses in older neighborhoods, the main supply line running from the street to the house is made of lead; this once posed a health hazard. Today, however, municipalities with lead pipes often add a trace amount of phosphate to the water, which coats the inside of the pipes and virtually eliminates leaching of lead into the water. If you are concerned about lead in your water, check with your local water supplier.

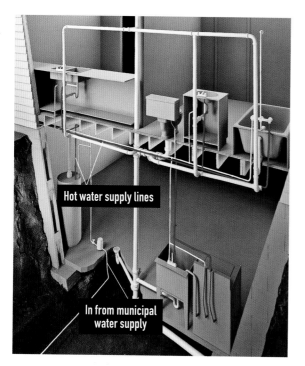

Hot water supply lines

In from municipal water supply

Cold water supply lines

Drain-Waste-Vent System

Drain pipes use gravity to carry waste water away from fixtures, appliances, and other drain openings. This waste water is carried out of the house to a municipal sewer system or septic tank.

Newer drain pipes are plastic. In an older home, drain pipes may be cast iron, galvanized steel, copper, or lead. Because they are not part of the supply system, lead drain pipes pose no health hazard. However, lead pipes are no longer manufactured for home plumbing systems.

Drain pipes have diameters ranging from 1¼" to 4". These large diameters allow waste to pass through efficiently.

Traps are an important part of the drain system. These curved sections of drain pipe hold standing water, and they are usually found immediately after the drain tailpiece in the drain opening. The standing water of a trap prevents sewer gases from backing up into the home. Each time a drain is used, the standing trap water is flushed away and is replaced by new water.

In order to work properly, the drain system requires air. Air allows waste water to flow freely down drain pipes.

To allow air into the drain system, drain pipes are connected to vent pipes. All drain systems must include vents, and the entire system is called the drain-waste-vent (DWV) system. One or more vent stacks, located on the roof, provide the air needed for the DWV system to work.

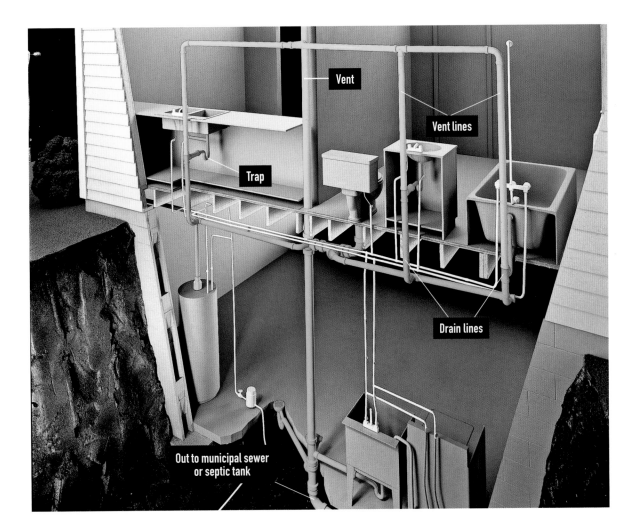

PLUMBING TOOLS

Many plumbing projects and repairs can be completed with basic hand tools you probably already own. Adding a few simple plumbing tools will prepare you for all the projects in this book. Specialty tools, such as a snap cutter or appliance dolly, are available at rental centers. When buying tools, invest in quality products.

Always care for tools properly. Clean tools after using them, wiping them free of dirt and dust with a soft rag. Prevent rust on metal tools by wiping them with a rag dipped in household oil. If a metal tool gets wet, dry it immediately, and then wipe it with an oiled rag. Keep toolboxes and cabinets organized. Make sure all tools are stored securely.

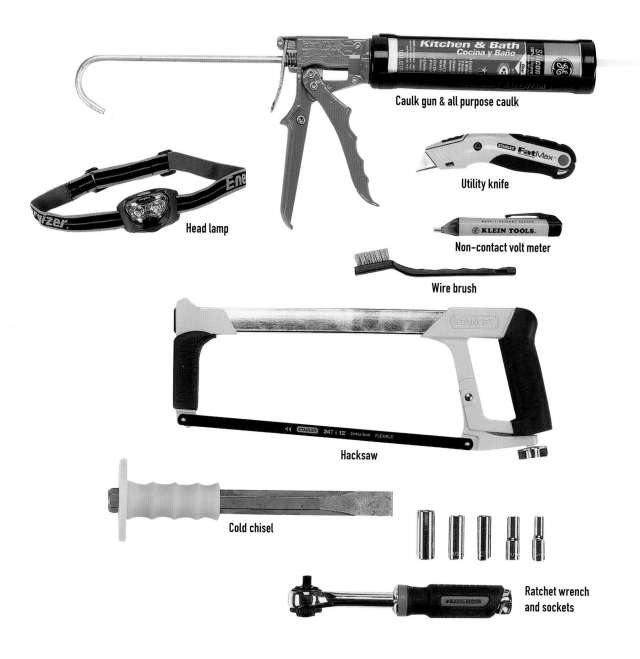

Caulk gun & all purpose caulk

Utility knife

Non-contact volt meter

Head lamp

Wire brush

Hacksaw

Cold chisel

Ratchet wrench and sockets

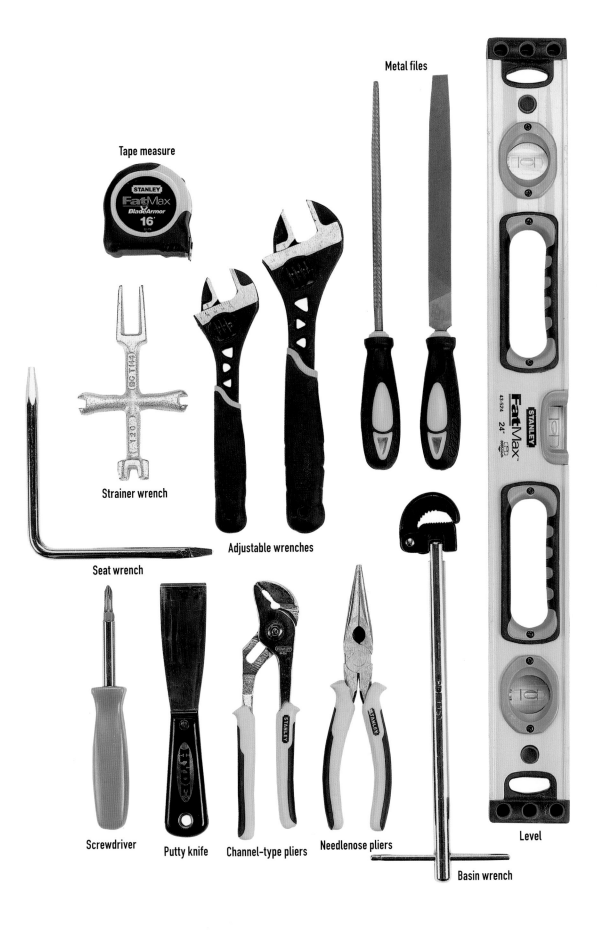

Tape measure

Metal files

Strainer wrench

Adjustable wrenches

Seat wrench

Screwdriver

Putty knife

Channel-type pliers

Needlenose pliers

Basin wrench

Level

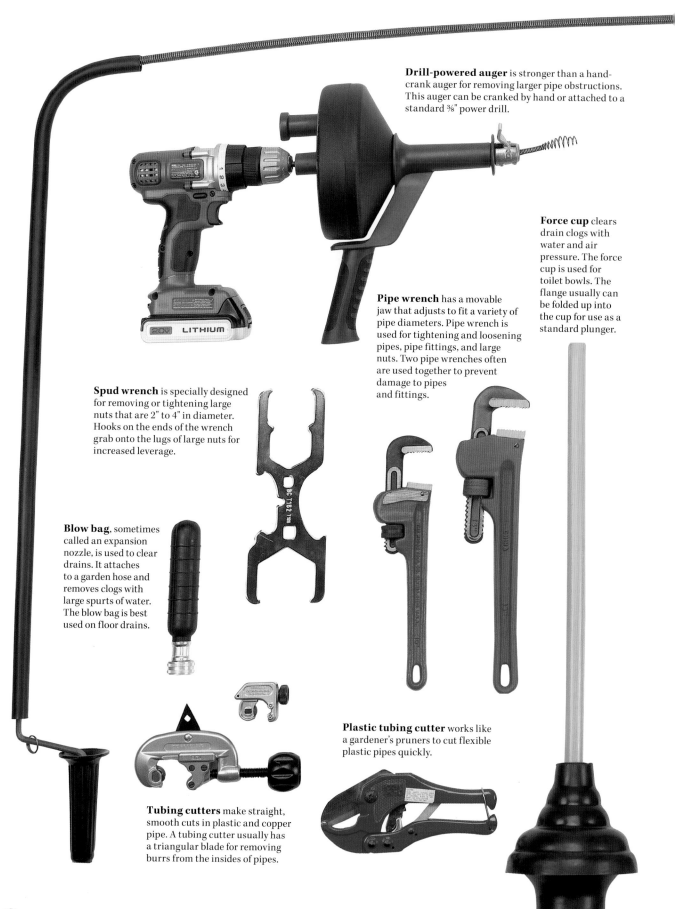

Drill-powered auger is stronger than a hand-crank auger for removing larger pipe obstructions. This auger can be cranked by hand or attached to a standard ⅜" power drill.

Force cup clears drain clogs with water and air pressure. The force cup is used for toilet bowls. The flange usually can be folded up into the cup for use as a standard plunger.

Pipe wrench has a movable jaw that adjusts to fit a variety of pipe diameters. Pipe wrench is used for tightening and loosening pipes, pipe fittings, and large nuts. Two pipe wrenches often are used together to prevent damage to pipes and fittings.

Spud wrench is specially designed for removing or tightening large nuts that are 2" to 4" in diameter. Hooks on the ends of the wrench grab onto the lugs of large nuts for increased leverage.

Blow bag, sometimes called an expansion nozzle, is used to clear drains. It attaches to a garden hose and removes clogs with large spurts of water. The blow bag is best used on floor drains.

Plastic tubing cutter works like a gardener's pruners to cut flexible plastic pipes quickly.

Tubing cutters make straight, smooth cuts in plastic and copper pipe. A tubing cutter usually has a triangular blade for removing burrs from the insides of pipes.

Closet auger is used to clear toilet clogs. It is a slender tube with a crank handle on one end of a flexible auger cable. A special bend in the tube allows the auger to be positioned in the bottom of the toilet bowl. The bend is usually protected with a rubber sleeve to prevent scratching the toilet.

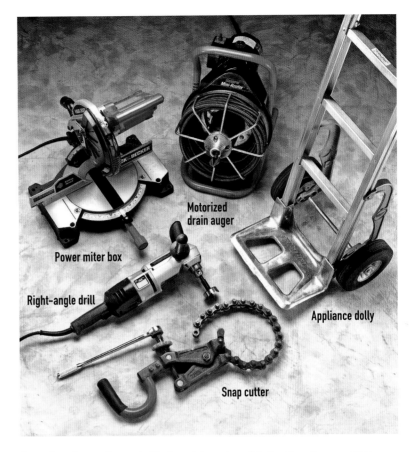

Power miter box

Right-angle drill

Motorized drain auger

Appliance dolly

Snap cutter

MAPP torch (left) is used for soldering fittings to copper pipes. Light the torch quickly and safely using a spark lighter.

Rental tools may be needed for large jobs and special situations. A power miter saw makes fast, accurate cuts in a wide variety of materials, including plastic pipes. A motorized drain auger clears tree roots from sewer service lines. Use an appliance dolly to move heavy objects like water heaters. A snap cutter is designed to cut tough cast-iron pipes. The right-angle drill is useful for drilling holes in hard-to-reach areas.

Flame-resistant pad helps keep wood and other underlying materials safe from the torch's flame.

Power drills and bits

Reciprocating saw

Power hand tools can make any job faster, easier, and safer. Cordless power tools offer added convenience. Use a cordless ⅜" power drill for virtually any drilling task.

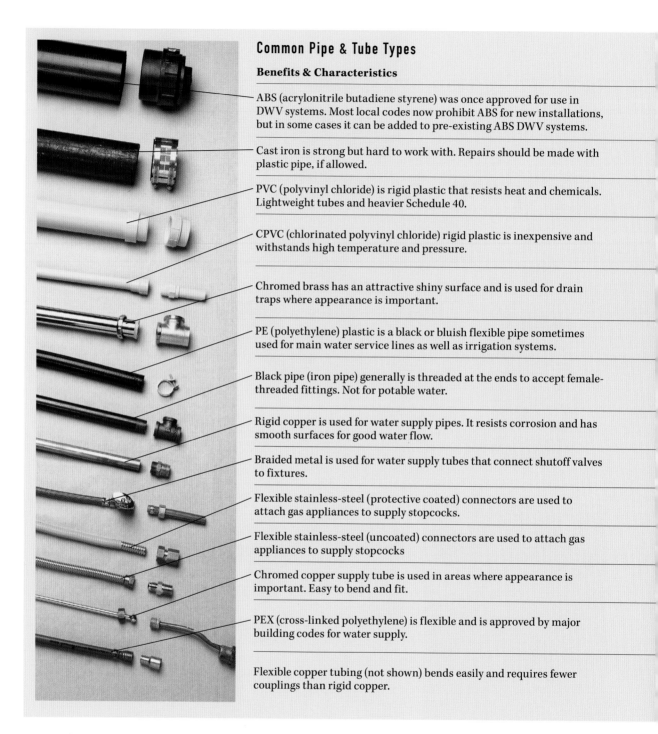

Common Pipe & Tube Types

Benefits & Characteristics

ABS (acrylonitrile butadiene styrene) was once approved for use in DWV systems. Most local codes now prohibit ABS for new installations, but in some cases it can be added to pre-existing ABS DWV systems.

Cast iron is strong but hard to work with. Repairs should be made with plastic pipe, if allowed.

PVC (polyvinyl chloride) is rigid plastic that resists heat and chemicals. Lightweight tubes and heavier Schedule 40.

CPVC (chlorinated polyvinyl chloride) rigid plastic is inexpensive and withstands high temperature and pressure.

Chromed brass has an attractive shiny surface and is used for drain traps where appearance is important.

PE (polyethylene) plastic is a black or bluish flexible pipe sometimes used for main water service lines as well as irrigation systems.

Black pipe (iron pipe) generally is threaded at the ends to accept female-threaded fittings. Not for potable water.

Rigid copper is used for water supply pipes. It resists corrosion and has smooth surfaces for good water flow.

Braided metal is used for water supply tubes that connect shutoff valves to fixtures.

Flexible stainless-steel (protective coated) connectors are used to attach gas appliances to supply stopcocks.

Flexible stainless-steel (uncoated) connectors are used to attach gas appliances to supply stopcocks

Chromed copper supply tube is used in areas where appearance is important. Easy to bend and fit.

PEX (cross-linked polyethylene) is flexible and is approved by major building codes for water supply.

Flexible copper tubing (not shown) bends easily and requires fewer couplings than rigid copper.

Common Uses	Lengths	Diameters	Fitting Methods	Tools Used for Cutting
Pipes; drain traps	Sold by linear ft.	2", 3", 4"	Glue and plastic	Miter box or hacksaw
Main drain- waste- vent stack	5 ft., 10 ft.	3", 4"	Banded neoprene couplings	Snap cutter or hacksaw
Drain & vent pipes; drain traps	10 ft., 20 ft.; or sold by linear ft.	1¼", 1½", 2", 3", 4"	Solvent glue and/or plastic fittings	Tubing cutter, miter box, or hacksaw
Hot & cold water supply pipes	10 ft.	⅜", ½", ¾", 1"	Solvent glue and plastic fittings, or with compression fittings	Tubing cutter, miter box, or hacksaw
Valves & shutoffs; drain traps, supply risers	Lengths vary	1¼", ½", ¾", 1¼", 1½"	Compression fittings, or with metal solder	Tubing cutter, hacksaw, or reciprocating saw
Outdoor cold water supply pipes	Sold in coils of 25 to hundreds of ft.	¼" to 1"	Rigid PVC fittings and stainless steel hose	Ratchet-style plastic pipe cutter or miter saw
Gas supply pipe	Sold in lengths up to 10 ft.	¾", 1"	Threaded connectors	Hacksaw, power cutoff saw or reciprocating saw with bi-metal blade
Hot & cold water supply pipes	10 ft., 20 ft.; or sold by linear ft.	⅜", ½", ¾", 1"	Metal solder or compression fittings	Tubing cutter, hacksaw, or jigsaw
Supply tubes	12" or 20"	⅜"	Compression coupling or compression fittings	Do not cut
Gas ranges, dryers, water heaters	36" or 48"	⅝", ½" (OD)	Compression coupling	Do not cut
Gas ranges, dryers, water heaters	36" or 48"	⅝", ½" (OD)	Compression coupling	Do not cut
Supply tubing	12", 20", 30"	⅜"	Brass compression fittings	Tubing cutter or hacksaw
Water supply, tubing for radiant floors	Sold in coils of 25 ft. to hundreds of ft.	¼" to 1"	Crimp fittings	Tubing cutter
Gas supply; hot & cold water supply	30-ft., 60-ft. coils; or by ft.	¼", ⅜", ½", ¾", 1"	Brass flare fittings, solder, compression fittings	Tubing cutter or hacksaw

COPPER

Copper resists corrosion and has smooth surfaces that allow good water flow. Copper pipes are available in several diameters, but most home water supply systems use ½-inch or ¾-inch pipe. Copper pipe is manufactured in rigid and flexible forms.

Rigid copper, sometimes called hard copper, is approved for home water supply systems by all local codes. It comes in three wall-thickness grades: Types M, L, and K. Type M is the thinnest, the least expensive, and a good choice for do-it-yourself home plumbing.

Rigid Type L usually is required by code for commercial plumbing systems. Because it is strong and solders easily, Type L may be preferred by some professional plumbers and do-it-yourselfers for home use. Type K has the heaviest wall thickness and is used most often for underground water service lines.

Flexible copper, also called soft copper, comes in two wall-thickness grades: Types L and K. Both are approved for most home water supply systems, although flexible Type L copper is used primarily for gas service lines. Because it is bendable and will resist a mild frost, Type L may be installed as part of a water supply system in unheated indoor areas, like crawl spaces. Type K is used for underground water service lines.

A third form of copper, called DWV, is used for drain systems. Because most codes now allow low-cost plastic pipes for drain systems, DWV copper is seldom used.

Skillbuilder

Practice makes perfect. If you have never soldered copper pipe before, practice sweating a few fittings onto a short sections of pipe before beginning an installation. This will acquaint you with process in a low stress setting.

Copper pipes are connected with soldered, compression, or flare fittings (see chart below). Always follow your local code for the correct types of pipes and fittings allowed in your area.

Soldered fittings, also called sweat fittings, often are used to join copper pipes. Correctly soldered fittings are strong and trouble-free. Copper pipe can also be joined with compression fittings or flare fittings. See chart below.

Copper Pipe & Fitting Chart

Fitting Method	Type M	Rigid Copper Type L	Type K	Flexible Copper Type L	Type K	General Comments
Soldered	Yes	Yes	Yes	Yes	Yes	Inexpensive, strong, and trouble-free fitting method. Requires some skill.
Compression	Yes	Not Applicable		No	No	Makes repairs and replacement easy. More expensive than solder. Best used on flexible copper.
Flare	No	No	Yes	Yes	Yes	Use only with flexible copper pipes. Usually used as a gas-line fitting. Requires some skill.

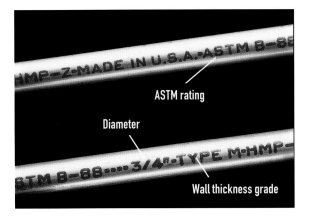

Grade stamp information includes the pipe diameter, the wall-thickness grade, and a stamp of approval from the ASTM (American Society for Testing and Materials). Type M pipe is identified by red lettering, Type L by blue lettering.

Bend flexible copper pipe with a coil-spring tubing bender to avoid kinks. Select a bender that matches the outside diameter of the pipe. Slip bender over pipe using a twisting motion. Bend pipe slowly until it reaches the correct angle, but not more than 90°.

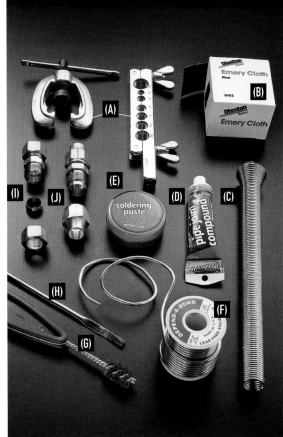

Specialty tools and materials for working with copper include: flaring tools (A), emery cloth (B), coil-spring tubing bender (C), pipe joint compound (D), soldering paste (flux) (E), lead-free solder (F), wire brush (G), flux brush (H), compression fitting (I), flare fitting (J).

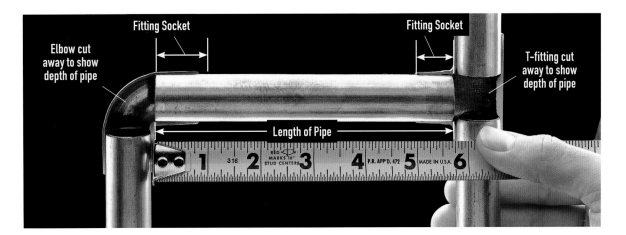

Find the length of copper pipe needed by measuring between the bottom of the copper fitting sockets (fittings shown in cutaway). Mark the length on the pipe with a felt-tipped pen.

Cut & Solder Copper

The best way to cut rigid and flexible copper pipe is with a tubing cutter. A tubing cutter makes a smooth, straight cut, an important first step toward making a watertight joint. Remove any metal burrs on the cut edges with a reaming tool or round file.

Copper can be cut with a hacksaw. A hacksaw is useful in tight areas where a tubing cutter will not fit. Take care to make a smooth, straight cut when cutting with a hacksaw.

A soldered pipe joint, also called a sweated joint, is made by heating a copper or brass fitting with a propane torch until the fitting is just hot enough to melt metal solder. The heat draws the solder into the gap between the fitting and pipe to form a watertight seal. A fitting that is overheated or unevenly heated will not draw in solder. Copper pipes and fittings must be clean and dry to form a watertight seal.

Tip

Never light a propane torch with a match. Ignition of the torch usually results in a small fire ball which will easily burn your fingers. A spark lighter keeps your hands and clothing at a safe distance.

Tips for Safe Soldering

Torch valve

Use caution when soldering copper. Pipes and fittings become very hot and must be allowed to cool before handling.

Prevent accidents by shutting off propane torch immediately after use. Make sure valve is closed completely.

Protect wood from the heat of the torch flame while soldering. Use an old cookie sheet, two sheets of 26-gauge metal, or a fiber shield, as shown.

Tools & Materials

Tubing cutter with reaming tip (or hacksaw and round file)
Wire brush
Flux brush
Propane torch

Spark lighter
Round file
Cloth
Adjustable wrench
Channel-type pliers
Copper pipe

Copper fittings
Emery cloth
Soldering paste (flux)
Sheet metal
Lead-free solder
Rag

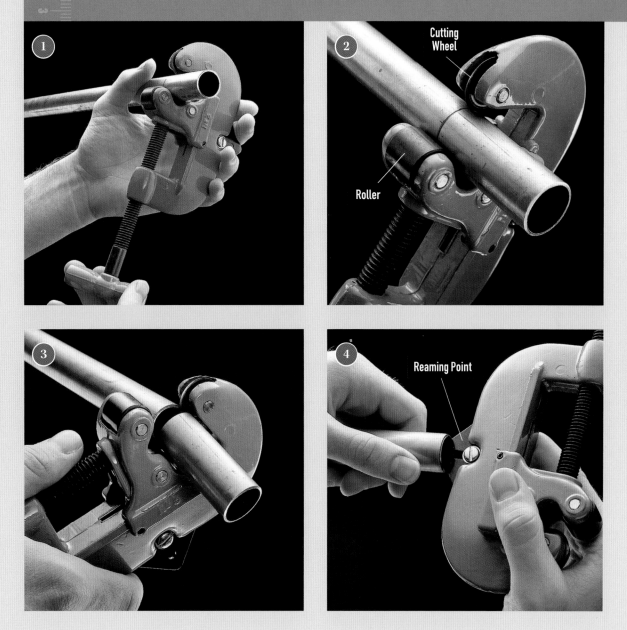

1   Place the tubing cutter over the pipe and tighten the handle so that the pipe rests on both rollers and the cutting wheel is on the marked line.

2   Turn the tubing cutter one rotation so that the cutting wheel scores a continuous straight line around the pipe.

3   Rotate the cutter in the opposite direction, tightening the handle slightly after every two rotations, until the cut is complete.

4   Remove sharp metal burrs from the inside edge of the cut pipe using the reaming point on the tubing cutter or a round file.

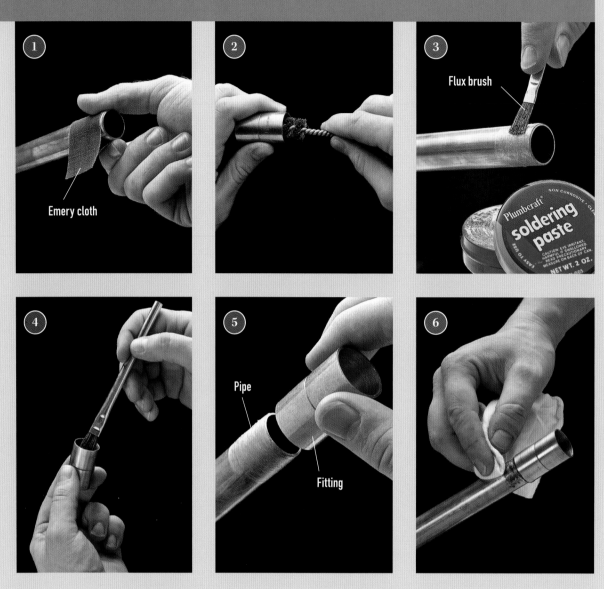

1 Clean the end of each pipe by sanding with emery cloth. Ends must be free of dirt and grease to ensure that the solder forms a good seal.

2 Clean the inside of each fitting by scouring with a wire brush or emery cloth.

3 Apply a thin layer of soldering paste (flux) to the end of each pipe, using a flux brush. Soldering paste should cover about 1" of pipe end.

4 Apply a thin layer of flux to the inside of the fitting.

5 Assemble each joint by inserting the pipe into the fitting so it is tight against the bottom of the fitting sockets. Twist each fitting slightly to spread soldering paste.

6 Use a clean dry cloth to remove excess flux before soldering the assembled fitting.

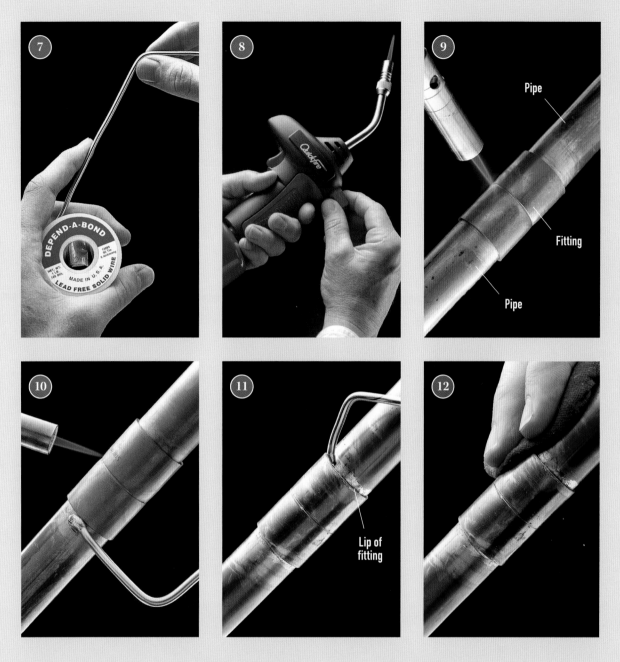

7   Prepare the wire solder by unwinding 8" to 10" of wire from the spool. Bend the first 2" of the wire to a 90° angle.

8   Open the gas valve and trigger the spark lighter to ignite the torch. Adjust the torch valve until the inner portion of the flame is 1" to 2" long.

9   Move the torch flame back and forth and around the pipe and the fitting to heat the area evenly.

10   Heat the other side of the copper fitting to ensure that heat is distributed evenly. Touch the solder to the pipe. The solder will melt when the pipe is at the right temperature.

11   When the solder melts, remove the torch and quickly push ½" to ¾" of solder into each joint. Capillary action fills the joint with liquid solder. A correctly soldered joint should show a thin bead of solder around the lips of the fitting.

12   Allow the joint to cool briefly, then wipe away excess solder with a dry rag. Caution: Pipes will be hot. If joints leak after water is turned on, disassemble and resolder.

SOLDERING BRASS VALVES

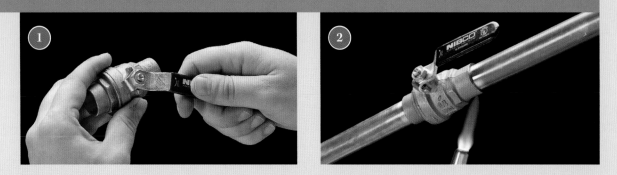

1   Valves should be fully open during all stages of the soldering process. If a valve has any plastic or rubber parts, remove them prior to soldering.

2   To prevent valve damage, quickly heat the pipe and the flanges of the valve, not the valve body. After soldering, cool the valve by spraying with water.

DISMANTLING SOLDERED JOINT BLADES

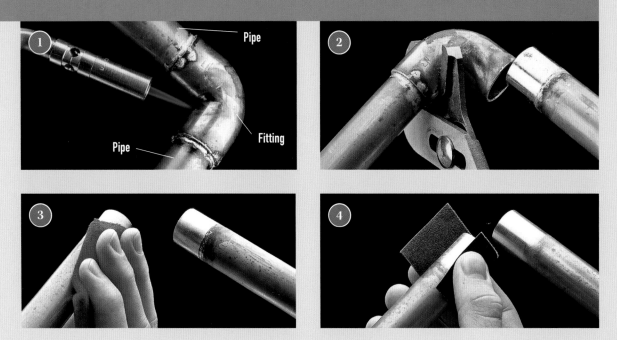

1   Turn off the water and drain the pipes by opening the highest and lowest faucets in the house. Light your torch. Hold the flame tip to the fitting until the solder becomes shiny and begins to melt.

2   Use channel-type pliers to separate the pipes from the fitting.

3   Remove old solder by heating the ends of the pipe with your torch. Use a dry rag to wipe away melted solder quickly. Caution: Pipes will be hot.

4   Use emery cloth to polish the ends of the pipe down to bare metal. Never reuse fittings.

Push fittings make water supply connections about as easy as possible. They are expensive, so you won't want to use them for all connections on a large installation.

But even professional plumbers use them in tight spots where sweating or welding would be difficult. They are also an ideal material for making a quick repair.

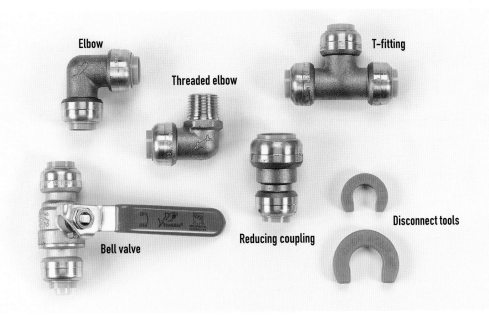

Push fittings are available as couplings tees, elbows, and even shutoff valves. They connect to hard copper, CPVC, and PEX pipe, but not to PVC or galvanized or black steel pipe. In most areas they are approved for use inside covered walls.

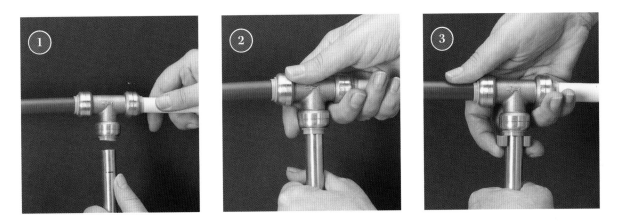

1. Cut the pipe square, and remove any burrs and rough edges. Draw a mark 1" from the cut end.

2. Push the pipe into the fitting an inch or so until you hear it click. Tug to make sure you have a strong connection. It may not seem like it, but the connection is indeed watertight and durable. You may rotate it to the desired position.

3. To remove a pipe from a push fitting, slip the disconnect tool over the pipe, slide it over the fitting, and press against the fitting's release collar as you pull the pipe out.

RIGID PLASTIC PIPE

Cut rigid ABS, PVC, or CPVC plastic pipes with a tubing cutter or with any saw. Cuts must be straight to ensure watertight joints.

Rigid plastics are joined with plastic fittings and solvent glue. Use a solvent glue that is made for the type of plastic pipe you are installing. For example, do not use ABS solvent on PVC pipe. Some solvent glues, called "all-purpose" or "universal" solvents, may be used on all types of plastic pipe.

Solvent glue hardens in about 30 seconds, so test-fit all plastic pipes and fittings before gluing the first joint. For best results, the surfaces of plastic pipes and fittings should be dulled with emery cloth and liquid primer before they are joined.

Liquid solvent glues and primers are toxic and flammable. Provide adequate ventilation when fitting plastics, and store the products away from any source of heat.

Plastic grip fittings can be used to join rigid or flexible plastic pipes to copper plumbing pipes.

Skillbuilder

Before you begin a plastic pipe project, take some time to practice the priming and cementing process. Once you apply the cement, you have a limited time to assemble. Buy some pipe and fittings and practice the solvent-gluing steps. It might seem expensive, but you need to understand the materials before undertaking a project.

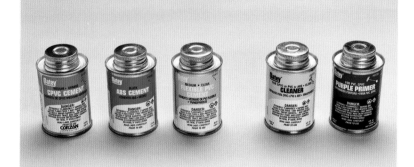

Primer and solvent glue are specific to the plumbing material being used. Do not use all-purpose or multi-purpose products. Light to medium body glues are appropriate for DIYers as they allow the longest working time and are easiest to use. When working with large pipe, 3 or 4 inches in diameter, buy a large-size can of cement, which has a larger dauber. If you use the small dauber (which comes with the small can), you may need to apply twice, which will slow you down and make connections difficult. (The smaller can of primer is fine for any other size pipe, since there's no rush in applying primer.) Cement (though not primer) goes bad in the can within a month or two after opening, so you may need to buy a new can for a new project.

Tools & Materials

Tape measure
Felt-tipped pen
Tubing cutter (or miter box
    or hacksaw)
Utility knife
Channel-type pliers
Gloves
Plastic pipe
Fittings
Emery cloth
Plastic pipe primer
Solvent glue
Rag
Petroleum jelly

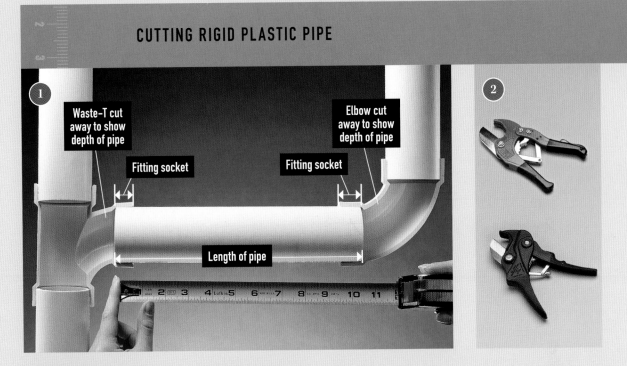

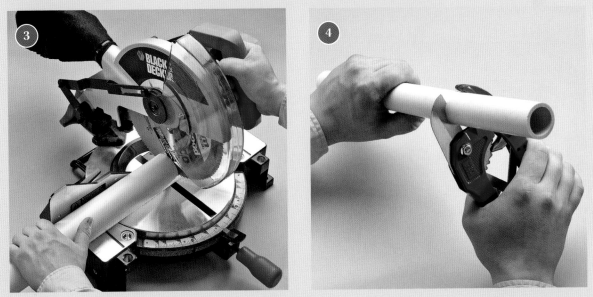

1   Find the length of plastic pipe needed by measuring between the bottoms of the fitting sockets (fittings shown in cutaway). Mark the length on the pipe with a felt-tipped pen.

2   Plastic tubing cutters do a fast, neat job of cutting. You'll probably have to go to a professional plumbing supply store to find one, however. They are not interchangeable with metal tubing cutters.

3   The best cutting tool for plastic pipe is a power miter saw with a fine tooth woodworking blade or a plastic-specific blade.

4   A ratcheting plastic-pipe cutter can cut smaller diameter PVC and CPVC pipe in a real hurry. If you are plumbing a whole house you may want to consider investing in one. They also are sold only at plumbing supply stores.

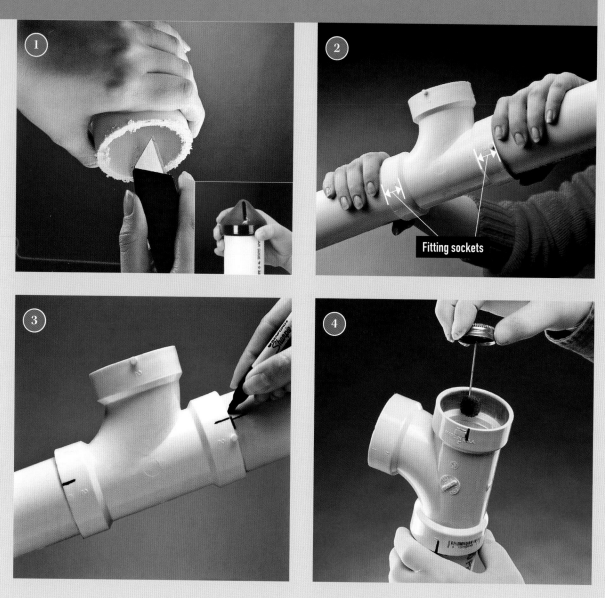

1   Remove rough burrs on cut ends of plastic pipe using a utility knife or deburring tool (inset).

2   Test-fit all pipes and fittings. Pipes should fit tightly against the bottom of the fitting sockets.

3   Mark the depth of the fitting sockets on the pipes. Take pipes apart. Clean the ends of the pipes and fitting sockets with emery cloth.

4   Apply a light coat of plastic pipe primer to the ends of the pipes and to the insides of the fitting sockets. Primer dulls glossy surfaces and ensures a good seal.

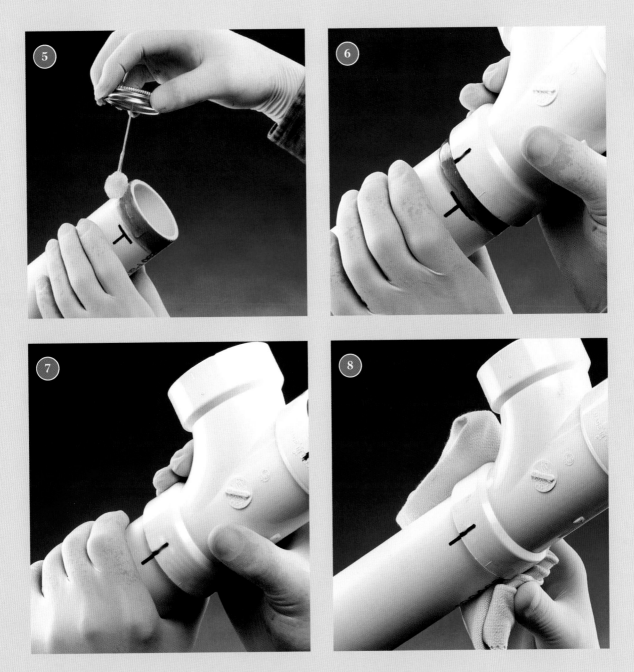

5   Solvent-glue each joint by applying a thick coat of solvent glue to the end of the pipe. Apply a thin coat of solvent glue to the inside surface of the fitting socket. Work quickly: solvent glue hardens in about 30 seconds.

6   Quickly position the pipe and fitting so that the alignment marks are offset by about 2". Force the pipe into the fitting until the end fits flush against the bottom of the socket.

7   Spread solvent by twisting the pipe until the marks are aligned. Hold the pipe in place for about 20 seconds to prevent the joint from slipping.

8   Wipe away excess solvent glue with a rag. Do not disturb the joint for 30 minutes after gluing.

OUTDOOR FLEXIBLE PLASTIC PIPE

Flexible PE (polyethylene) pipe is used for underground cold water lines. Very inexpensive, PE pipe is commonly used for automatic lawn sprinkler systems and for extending cold water supply to utility sinks in detached garages and sheds.

Unlike other plastics, PE is not solvent-glued, but is joined using "barbed" rigid PVC fittings and stainless-steel hose clamps. In cold climates, outdoor plumbing lines should be shut off and drained for winter.

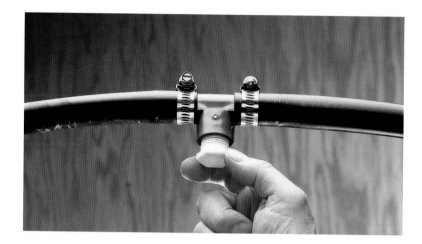

Connect lengths of PE pipe with a barbed PVC fitting. Secure the connection with stainless steel hose clamps.

Tools & Materials

Tape measure
Tubing cutter
Screwdriver or wrench
Pipe joint compound
Flexible pipe
Fittings
Hose clamps
Utility knife

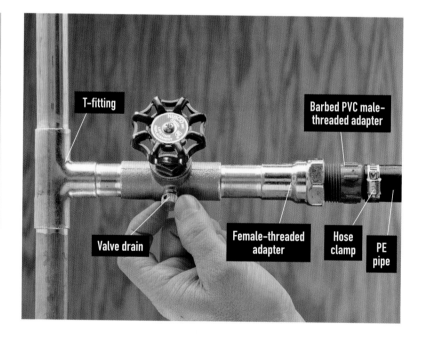

Connect PE pipe to an existing cold water supply pipe by splicing in a T-fitting to the copper pipe and attaching a drain-and-waste shutoff valve and a female-threaded adapter. Screw a barbed PVC male-threaded adapter into the copper fitting, then attach the PE pipe. The drain-and-waste valve allows you to blow the PE line free of water when winterizing the system.

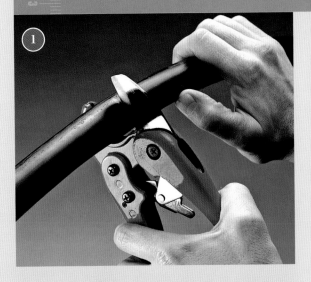

Option

To ensure a tighter fit, dab some pipe joint compound onto the barbs so they are easier to slide into the flexible plastic pipe. Apply pipe joint compound to the barbed ends of the T-fitting. Work each end of PE pipe over the barbed portions of the fitting and into position.

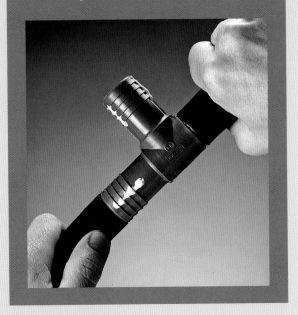

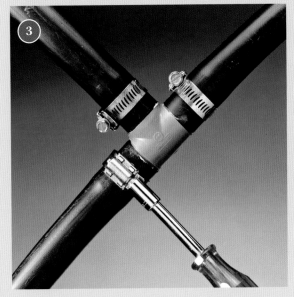

1   Cut flexible PE pipe with a plastic tubing cutter, or use a miter box or sharp knife. Remove any rough burrs with a utility knife.

2   Fit stainless-steel hose clamps over the ends of the flexible pipes being joined.

3   Slide the band clamps over the joint ends. Hand tighten each clamp with a screwdriver or wrench.

CROSS-LINKED POLYETHYLENE (PEX)

Cross-linked polyethylene (PEX) is growing quickly in acceptance as a supply pipe for residential plumbing. It's not hard to understand why. Developed in the 1960s but relatively new to the United States, this supply pipe combines the ease of use of flexible tubing with the durability of rigid pipe. It can withstand a wide temperature range (from subfreezing to 180°F); it is inexpensive; and it's quieter than rigid supply pipe.

PEX is flexible plastic (polyethylene, or PE) tubing that's reinforced by a chemical reaction that creates long fibers to increase the strength of the material. It has been allowed by code in Europe and the southern United States for many years, but has won approval for residential supply use in most major plumbing codes only recently. It's frequently used in manufactured housing and recreational vehicles and in radiant heating systems. Because it is so flexible, PEX can easily be bent to follow corners and make other changes in direction. From the water main and heater, it is connected into manifold fittings that redistribute the water in much the same manner as a lawn irrigation system.

For standard residential installations, PEX can be joined with very simple fittings and tools. Unions are generally made with a crimping tool and a crimping ring. You simply insert the ends of the pipe you're joining into the ring, then clamp down on the ring with the crimping tool. PEX pipe, tools, and fittings can be purchased from most wholesale plumbing suppliers and at many home centers. Coils of PEX are sold in several diameters, from ¼-inch to 1-inch. PEX tubing and fittings from different manufacturers are not interchangeable. Any warranty coverage will be voided if products are mixed.

Tools & Materials

Tape measure	Manifolds
Felt-tipped pen	Protector plates
Full-circle crimping tool	PEX fittings
Go/no-go gauge	Utility knife
Tubing cutter	Plastic hangers
PEX pipe	Crimp ring

PEX pipe is a relatively new water supply material that's growing in popularity in part because it can be installed with simple mechanical connections.

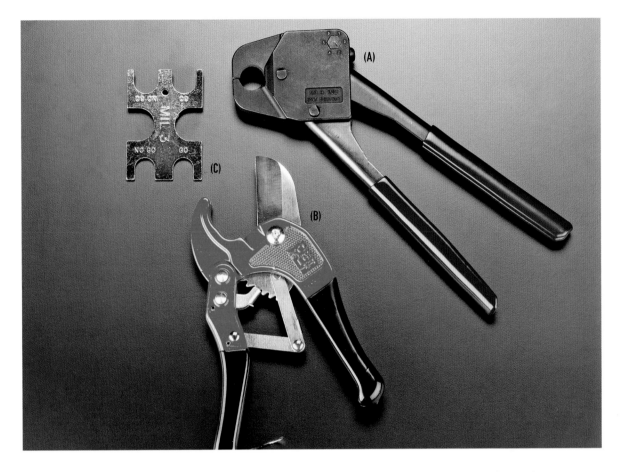

Specialty tools for installing PEX are available wherever PEX is sold. The basic set includes a full-circle crimping tool (A), a tubing cutter (B), and a go/no-go gauge (C) to test connections after they've been crimped. Competing manufacturers make several types of fittings, with proprietary tools that only work with their fittings. The tools and fittings you use may differ from those shown on these pages.

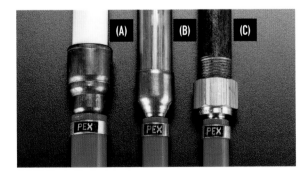

PEX is connected to other water supply materials with transition fittings, including CPVC-to-PEX (A), copper-to-PEX (B), and iron-to-PEX (C).

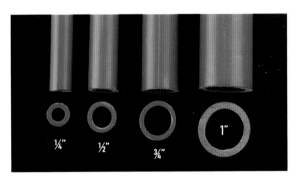

Generally, you should use the same diameter PEX as is specified for rigid supply tubing, but in some "home run" installations (see page 36) you can use ⅜" PEX where ½" rigid copper would normally be used.

PEX INSTALLATION

Check with your local plumbing inspector to verify that PEX is allowed in your municipality. PEX has been endorsed by all major plumbing codes in North America, but your municipality may still be using an older set of codes. Follow the guidelines below when installing PEX:

- Do not install PEX in above-ground exterior applications because it degrades quickly from UV exposure.
- Do not use PEX for gas lines.
- Do not use plastic solvents or petroleum-based products with PEX (they can dissolve the plastic).
- Keep PEX at least 12 inches away from recessed light fixtures and other potential sources of high heat.
- Do not attach PEX directly to a water heater. Make connections at the heater with metallic tubing (either flexible water-heater connector tubing or rigid copper) at least 18 inches long; then join it to PEX with a transition fitting.
- Do not install PEX in areas where there is a possibility of mechanical damage or puncture. Always fasten protective plates to wall studs that house PEX.
- Always leave some slack in installed PEX lines to allow for contraction and in case you need to cut off a bad crimp.
- Use the same minimum branch and distribution supply-pipe dimensions for PEX that you'd use for copper or CPVC, according to your local plumbing codes.
- You can use push fittings to join PEX to itself or to CPVC or copper.

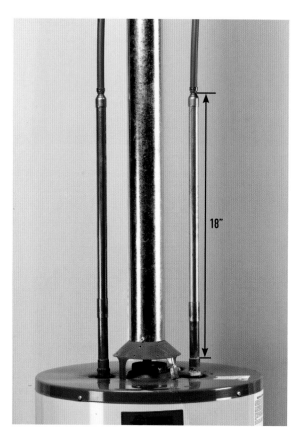

Do not connect PEX directly to a water heater. Use metal connector tubes. Solder the connector tubes to the water heater before attaching PEX. Never solder metal tubing that is already connected to PEX lines.

Bundle PEX together with plastic ties when running pipe through wall cavities. PEX can contract slightly, so leave some slack in the lines.

Buying PEX

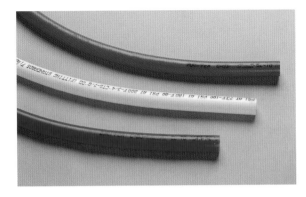

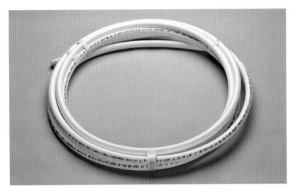

Color coding is a practice many PEX manufacturers have embraced to make identification easier. Because the material is identical except for the color, you can buy only one color (red is more common) and use it for both hot and cold supply lines.

PEX combines the flexibility of plastic tubing with the durability of rigid supply pipe. It is sold in coils of common supply-pipe diameters.

The PEX Advantage

PEX supply tubing offers a number of advantages over traditional rigid supply tubing:

- Easy to install. PEX does not require coupling joints for long runs or elbows and sweeps for turns. The mechanical connections do not require solvents or soldering.
- Easy to transport. Large coils are lightweight and much easier to move around than 10-ft. lengths of pipe.
- Good insulation. The PEX material has better thermal properties than copper for lessened heat loss.

- Quiet. PEX will not rattle or clang from trapped air or kinetic energy.
- Good for retrofit jobs. PEX is easier to snake through walls than rigid supply tubing and is compatible with copper, PVC, or iron supply systems if the correct transition fittings are used. If your metal supply tubes are used to ground your electrical system, you'll need to provide a jumper if PEX is installed in midrun. Check with a plumber or electrician.
- Freeze resistant. PEX retains some flexibility in sub-freezing conditions and is less likely to be damaged than rigid pipe, but it is not frostproof.

General Codes for PEX

PEX has been endorsed for residential use by all major building codes, although some municipal codes may be more restrictive. The specific design standards may also vary, but here are some general rules:

- For PEX, maximum horizontal support spacing is 32" and maximum vertical support spacing is 10 ft.

- Maximum length of individual distribution lines is 60 ft.
- PEX is designed to withstand 210°F water for up to 48 hours. For ongoing use, most PEX is rated for 180 degree water up to 100 pounds per square inch of pressure.
- Directional changes of more than 90 degrees require a guide fitting (see page 45).

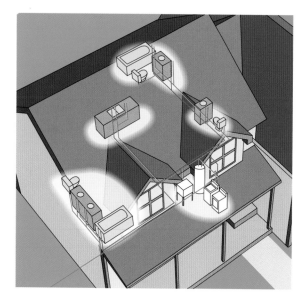

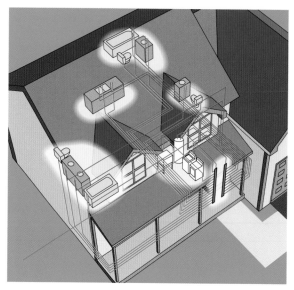

Trunk-and-branch systems are configured in much the same way as a traditional rigid copper or PVC supply systems. A main supply line (the trunk line) carries water to all of the outlets via smaller branch lines that tie into the trunk and serve a few outlets in a common location.

Home run systems rely on one or two central manifolds to distribute the hot and cold water very efficiently. Eliminating the branch fittings allows you to use thinner supply pipe in some situations.

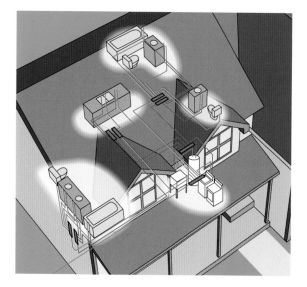

Remote manifold systems are a hybrid between traditional trunk-and-branch systems and home run systems. Instead of relying on just one or two manifolds, they employ several smaller manifolds downline from a larger manifold. Each smaller manifold services a group of fixtures, as in a bathroom or kitchen.

Choosing a PEX System

- For maximum single-fixture water pressure: Trunk and branch
- For economy of materials: Trunk and branch or remote manifold
- For minimal wait times for hot water (single fixture): Home run
- For minimal wait times for hot water (multiple fixtures used at same approximate time): Trunk and branch or remote manifold
- For ease of shutoff control: Home run
- For lowest number of fittings and joints: Home run

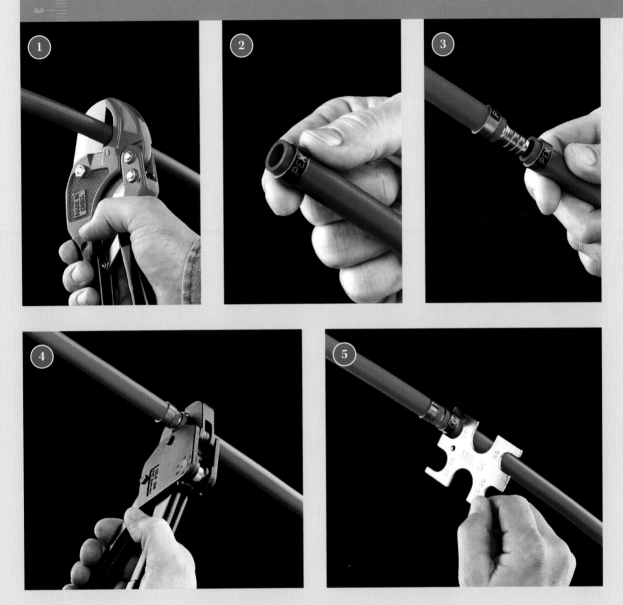

1   Cut the pipe to length, making sure to leave enough extra material so the line will have a small amount of slack once the connections are made. A straight, clean cut is very important. For best results, use a tubing cutter.

2   Inspect the cut end to make sure it is clean and smooth. If necessary, deburr the end of the pipe with a sharp utility knife. Slip a crimp ring over the end.

3   Insert the barbed end of the fitting into the pipe until it is snug against the cut edges. Position the crimp ring so it is ⅛" to ¼" from the end of the pipe, covering the barbed end of the fitting. Pinch the fitting to hold it in place.

4   Align the jaws of a full-circle crimping tool over the crimp ring and squeeze the handles together to apply strong, even pressure to the ring.

5   Test the connection to make sure it is mechanically acceptable, using a go/no-go gauge. If the ring does not fit into the gauge properly, cut the pipe near the connection and try again.

GALVANIZED STEEL

Galvanized steel pipe is often found in older homes, where it is used for water supply and small drain lines. It can be identified by the zinc coating that gives it a silver color and by the threaded fittings used to connect pipes.

Glavanized steel pipes and fittings will corrode with age and eventually must be replaced. Low water pressure may be a sign that the insides of galvanized pipes have a buildup of rust or other minerals. Blockage usually occurs in elbow fittings. Never try to clean the insides of galvanized steel pipes. Instead, remove and replace them as soon as possible.

Glavanized steel pipe and fittings are available at hardware stores and home improvement centers. Always specify the interior diameter (I.D.) when purchasing galvanized pipes and fittings. Pre-threaded pipes, called nipples, are available in lengths from 1 inch to 1 foot. If you need a longer length, have the store cut and thread the pipe to your dimensions.

Old galvanized steel can be difficult to repair. Fittings are often rusted in place, and what seems like a small job may become a large project. For example, cutting apart a section of pipe to replace a leaky fitting may reveal that adjacent pipes are also in need of replacement. If your job takes an unexpected amount of time, you can cap off any open lines and restore water to the rest of your house.

Tools & Materials	
Tape measure	Wire brush
Reciprocating saw with	Nipples
metal-cutting blade	End caps
or a hacksaw	Union fitting
Pipe wrenches	Pipe joint compound
Propane torch	Replacement fittings
	(if needed)

Before you begin a repair, have on hand nipples and end caps that match your pipes.

Taking apart a system of galvanized steel pipes and fittings is time-consuming. Disassembly must start at the end of a pipe run, and each piece must be unscrewed before the next piece can be removed. Reaching the middle of a run to replace a section of pipe can be a long and tedious job. Instead, use a special three-piece fitting called a union. A union makes it possible to remove a section of pipe or a fitting without having to take the entire system apart.

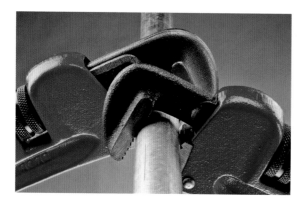

Galvanized pipe was installed in homes for both gas and water supply pipes until the middle part of the last century. Although it is not used for new installations today, it can still be repaired easily using simple tools and techniques.

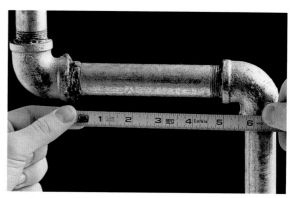

Measure the old pipe. Include ½" at each end for the threaded portion of the pipe inside fitting. Bring overall measurement to the store when shopping for replacement parts.

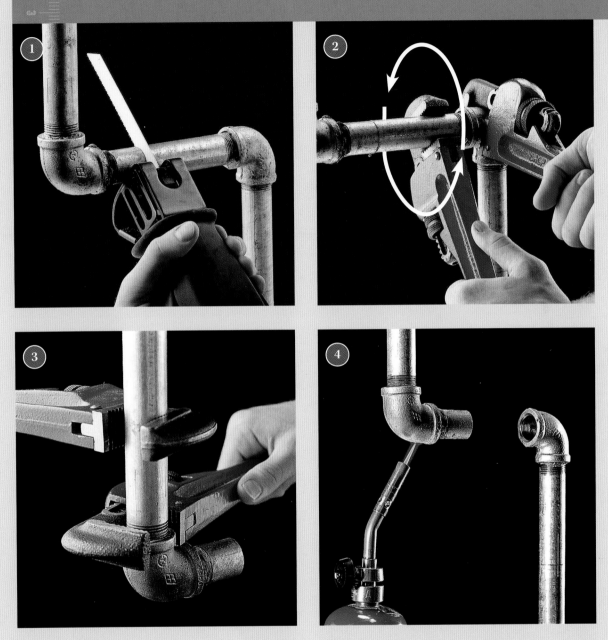

1   Cut through galvanized steel pipe with a reciprocating saw and a metal-cutting blade or with a hacksaw.

2   Hold the fitting with one pipe wrench, and use another wrench to remove the old pipe. The jaws of the wrenches should face opposite directions. Always move the wrench handle toward the jaw opening.

3   Remove any corroded fittings using two pipe wrenches. With the jaws facing in opposite directions, use one wrench to turn fitting and the other to hold the pipe. Clean the pipe threads with a wire brush.

4   Heat stubborn fittings with a torch to make them easier to remove. Apply the flame for 5 to 10 seconds. Protect wood and other flammable materials from heat using a double layer of sheet metal.

continued

PIPE FITTINGS

Use the photos on these pages to identify the plumbing fittings specified in the project how-to directions found in this book. Each fitting shown is available in a variety of sizes to match your needs. Always use fittings made from the same material as your pipes.

Pipe fittings come in a variety of shapes to serve different functions within the plumbing system. DWV fittings include:

Vents: In general, the fittings used to connect vent pipes have very sharp bends with no sweep. Vent fittings include the vent T and vent 90º elbow. Standard drain pipe fittings can also be used to join vent pipes.

Horizontal-to-vertical drains: To change directions in a drain pipe from the horizontal to the vertical, use fittings with a noticeable sweep. Standard fittings for this use include waste T-fittings and 90º elbows. Y-fittings and 45º and 22 1/2º elbows can also be used for this purpose.

Vertical-to-horizontal drains: To change directions from the vertical to the horizontal, use fittings with a very pronounced, gradual sweep. Common fittings for this purpose include the long-radius T-Y-fitting and some Y-fittings with 45º elbows.

Horizontal offsets in drains: Y-fittings, 45º elbows, 22 1/2º elbows, and long sweep 90º elbows are used when changing directions in horizontal pipe runs. Whenever possible, horizontal drain pipes should use gradual, sweeping bends rather than sharp turns.

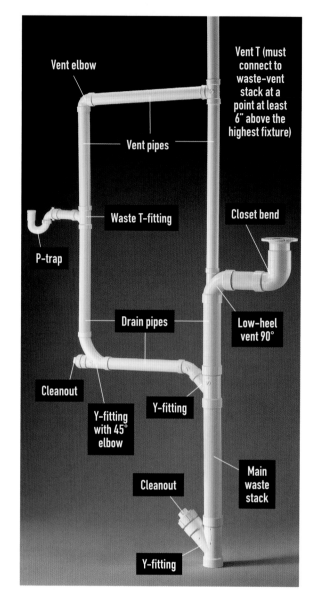

Basic DWV tree shows the correct orientation of drain and vent fittings in a plumbing system. Bends in the vent pipes can be very sharp, but drain pipes should use fittings with a noticeable sweep. Fittings used to direct falling waste water from a vertical to a horizontal pipe should have bends that are even more sweeping. Your local plumbing code may require that you install cleanout fittings where vertical drain pipes meet horizontal runs.

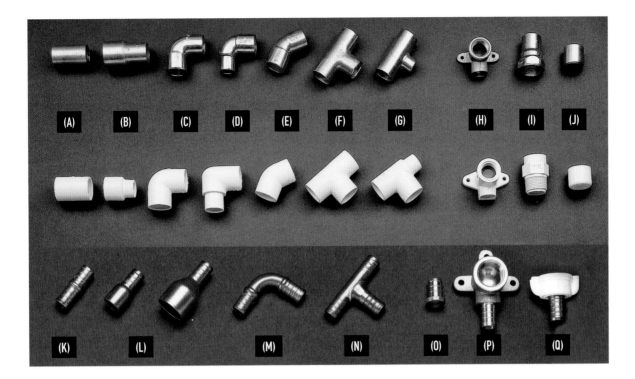

Water supply fittings are available for copper (top), CPVC plastic (center), and PEX (bottom). Fittings for CPVC and copper are available in many shapes, including: unions (A), reducers (B), 90° elbows (C), reducing elbows (D), 45° elbows (E), T-fittings (F), reducing T-fittings (G), drop-ear elbows (H), threaded adapters (I), and caps (J). Common PEX fittings (bottom) include unions (K), PEX-to-copper unions (L), 90° elbows (M), T-fittings (N), plugs (O), drop-ear elbows (P), and threaded adapters (Q). Easy-to-install push fittings are also available.

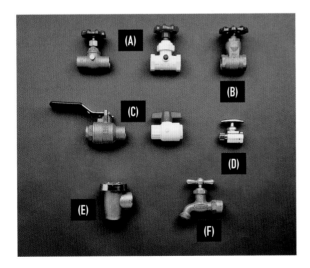

Water supply valves are available in brass or plastic and in a variety of styles, including: drain-and-waste valves (A), gate valve (B), full-bore ball valves (C), fixture shutoff valve (D), vacuum breaker (E), and hose bib (F).

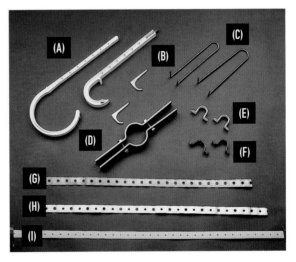

Support materials for pipes include: plastic pipe hangers (A), copper J-hooks (B), copper wire hangers (C), riser clamp (D), plastic pipe straps (E), copper pipe straps (F), flexible copper, steel, and plastic pipe strapping (G, H, I). Do not mix metal types when supporting metal pipes; use copper support materials for copper pipe, and steel for steel and cast-iron pipes.

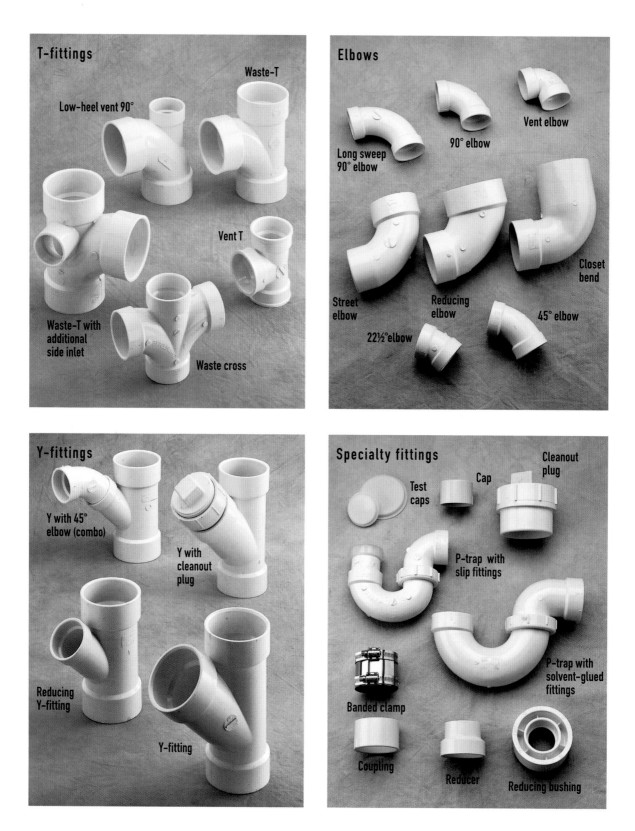

T-fittings

Waste-T

Low-heel vent 90°

Vent T

Waste-T with additional side inlet

Waste cross

Elbows

Long sweep 90° elbow

90° elbow

Vent elbow

Closet bend

Street elbow

Reducing elbow

22½° elbow

45° elbow

Y-fittings

Y with 45° elbow (combo)

Y with cleanout plug

Reducing Y-fitting

Y-fitting

Specialty fittings

Test caps

Cap

Cleanout plug

P-trap with slip fittings

Banded clamp

P-trap with solvent-glued fittings

Coupling

Reducer

Reducing bushing

Fittings for DWV pipes are available in many configurations, with openings ranging from 1¼" to 4" in diameter. When planning your project, buy plentiful numbers of DWV and water supply fittings from a reputable retailer with a good return policy. It is much more efficient to return leftover materials after you complete your project than it is to interrupt your work each time you need to shop for a missing fitting.

3" no-hub neoprene coupling

1½"-to-1¼" reducing transition

Connect plastic to cast iron with banded couplings. Rubber sleeves cover ends of pipes and ensure a watertight joint.

Make transitions in DWV pipes with rubber couplings. The two products shown here can be used to connect pipes of different materials, as well as same-material pipes that need a transition.

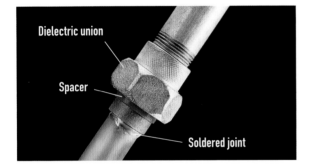

Dielectric union

Spacer

Soldered joint

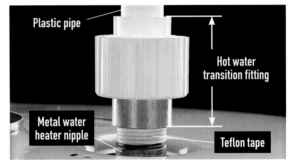

Plastic pipe

Hot water transition fitting

Metal water heater nipple

Teflon tape

Connect copper to galvanized steel with a dielectric union. A dielectric union is threaded onto iron pipe and is soldered to copper pipe. A dielectric union has a plastic spacer that prevents corrosion caused by an electrochemical reaction between dissimilar metals.

Connect metal hot water pipe to plastic with a hot water transition fitting that prevents leaks caused by different expansion rates of materials. Metal pipe threads are wrapped with Teflon tape. Plastic pipe is solvent-glued to fitting.

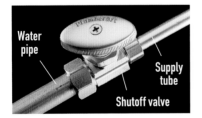

Water pipe

Supply tube

Shutoff valve

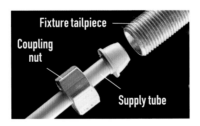

Fixture tailpiece

Coupling nut

Supply tube

Connect a water pipe to any fixture supply tube using a shutoff valve.

Connect any supply tube to a fixture tailpiece with a coupling nut. The coupling nut compresses the bell-shaped end of the supply tube against the fixture tailpiece.

Specialty supply fittings can be used to supply potable water fixtures such as icemakers and hot water dispensers. The John-Guest® Speed-Fit® fitting shown here is designed to connect to clear tubing or the manufacturer's proprietary plastic supply tubing.

SHUTOFF VALVES

Worn-out shutoff valves or supply tubes can cause water to leak underneath a sink or other fixture. First, try tightening the fittings with an adjustable wrench. If this does not fix the leak, replace the shutoff valves and supply tubes.

Shutoff valves are available in several fitting types. For copper pipes, valves with compression-type fittings are easiest to install. For plastic pipes, use grip-type valves. For galvanized steel pipes, use valves with female threads.

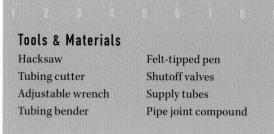

Tools & Materials

Hacksaw	Felt-tipped pen
Tubing cutter	Shutoff valves
Adjustable wrench	Supply tubes
Tubing bender	Pipe joint compound

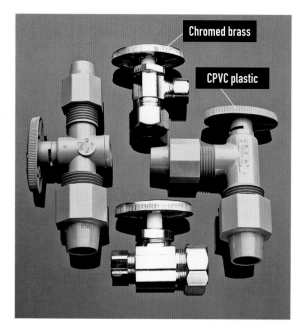

Shutoff valves allow you to shut off the water to an individual fixture so it can be repaired. They can be made from durable chromed brass or lightweight plastic. Shutoff valves come in ½" and ¾" diameters to match common water pipe sizes.

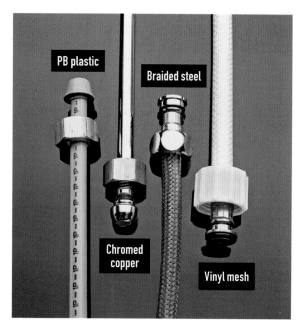

Supply tubes are used to connect water pipes to faucets, toilets, and other fixtures. They come in 12", 20", and 30" lengths. PB plastic and chromed copper tubes are inexpensive. Braided steel and vinyl mesh supply tubes are easy to install.

Tip
Older plumbing systems often were installed without fixture shutoff valves. When repairing or replacing plumbing fixtures, you may want to install shutoff valves if they are not already present.

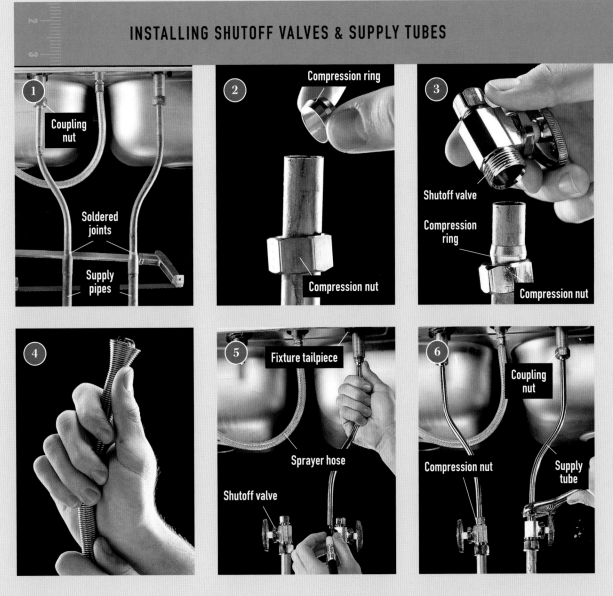

1   Turn off water at the main shutoff valve. Remove old supply pipes. If pipes are soldered copper, cut them off just below the soldered joint using a hacksaw or tubing cutter. Make sure the cuts are straight. Unscrew the coupling nuts and discard the old pipes.

2   Slide a compression nut and a compression ring over the copper water pipe. Threads of the nut should face the end of the pipe.

3   Apply pipe joint compound to the threads of the shutoff valve or compression nut. Screw the compression nut onto the shutoff valve and tighten with an adjustable wrench.

4   Bend chromed copper supply tube to reach from the tailpiece of the fixture to the shutoff valve using a tubing bender. Bend the tube slowly to avoid kinking the metal.

5   Position the supply tube between the fixture tailpiece and the shutoff valve, and mark the tube to length. Cut the supply tube with a tubing cutter (page 14).

6   Attach the bell-shaped end of the supply tube to the fixture tailpiece with a coupling nut, then attach the other end to the shutoff valve with compression ring and nut. Tighten all fittings with an adjustable wrench.

VALVES & HOSE BIBS

Valves make it possible to shut off water at any point in the supply system. If a pipe breaks or a plumbing fixture begins to leak, you can shut off water to the damaged area so that it can be repaired. A hose bib is a faucet with a threaded spout, often used to connect rubber utility or appliance hoses.

Valves and hose bibs leak when washers or seals wear out. Replacement parts can be found in the same universal washer kits used to repair compression faucets. Coat replacement washers with faucet grease to keep them soft and prevent cracking.

Tip

If you have the opportunity to replace a shutoff valve, install a ball valve, which is most reliable type.

Tools & Materials

Screwdriver
Adjustable wrench

Universal washer kit
Faucet grease

With the exception of chromed shutoff valves that are installed at individual fixtures (see previous pages), valves and hose bibs are heavy-duty fittings, usually with a brass body that are installed in-line to regulate water flow. Gate valves and globe valves are similar and are operated with a wheel-type handle that spins. Ball valves are operated with a handle much like a gas pipe stopcock and are considered by pros to be the most reliable. Hose bibs are spigots with a threaded end designed to accept a female hose coupling.

FIXING A LEAKY HOSE BIB

Packing nuts

Handle screw
Handle
Packing nut
Packing washer
Packing ring
Spindle
Stem washer
Stem screw

1   Turn off the water to the hose bib before beginning work. Remove the handle screw and lift off the handle. Unscrew the packing nut with an adjustable wrench.

2   Unscrew the spindle from the valve body. Remove the stem screw and replace the stem washer. Replace the packing washer and reassemble the valve.

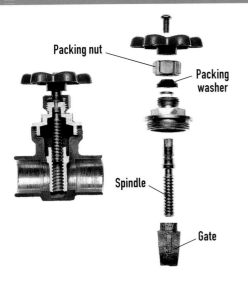

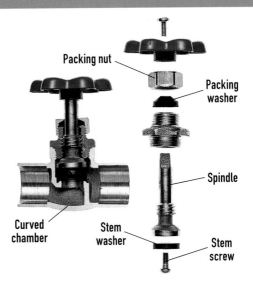

A gate valve has a movable brass wedge, or "gate," that screws up and down to control water flow. Gate valves may develop leaks around the handle. Repair leaks by replacing the packing washer or packing string found underneath the packing nut.

A globe valve has a curved chamber. Repair leaks around the handle by replacing the packing washer. If the valve does not fully stop water flow when closed, replace the stem washer.

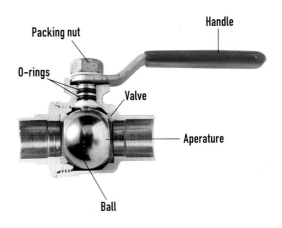

A shutoff valve controls water supply to one or more fixtures. A shutoff valve has a plastic spindle with a packing washer and a snap-on stem washer. Repair leaks around the handle by replacing the packing washer. If a valve does not fully stop water flow when closed, replace the stem washer. Shutoff valves with multiple outlets are available to supply several fixtures from a single supply.

A ball valve contains a metal ball with an aperture (or controlled hole) in the center. The ball is controlled by a handle. When the handle is turned the hole is positioned parallel to the valve (open) or perpendicular (closed).

COMPRESSION FITTINGS

Compression fittings are used to make connections that may need to be taken apart. Compression fittings are easy to disconnect and are often used to install supply tubes and fixture shutoff valves. Use compression fittings in places where it is unsafe or difficult to solder, such as in crawl spaces.

Compression fittings are used most often with flexible copper pipe. Flexible copper is soft enough to allow the compression ring to seat snugly, creating a watertight seal. Compression fittings also may be used to make connections with Type M rigid copper pipe.

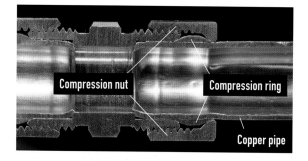

Compression fitting (shown in cutaway) shows how a threaded compression nut forms a seal by forcing the compression ring against the copper pipe. The compression ring is covered with pipe joint compound before assembling to ensure a perfect seal.

Tools & Materials

Felt-tipped pen
Tubing cutter or
   hacksaw
Adjustable wrenches

Brass compression
   fittings
Pipe joint compound or
   Teflon tape

ATTACHING SUPPLY TUBES TO FIXTURE SHUTOFF VALVES WITH COMPRESSION FITTING

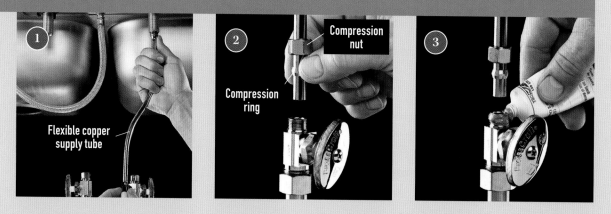

1 Bend the flexible copper supply tube and mark to length. Include ½" for the portion that will fit inside the valve. Cut the tube.

2 Slide the compression nut and then the compression ring over the end of the pipe. The threads of the nut should face the valve.

3 Apply a small amount of pipe joint compound to the threads. This lubricates the threads.

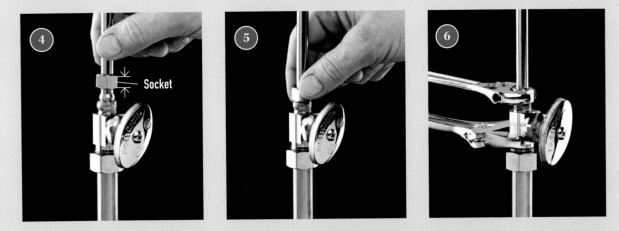

4   Insert the end of the pipe into the fitting so it fits flush against the bottom of the fitting socket.

5   Slide the compression ring and nut against the threads of the valve. Hand tighten the nut onto the valve.

6   Tighten the compression nut with adjustable wrenches. Do not overtighten. Turn on the water and watch for leaks. If the fitting leaks, tighten the nut gently.

JOINING TWO COPPER PIPES WITH A COMPRESSION UNION FITTING

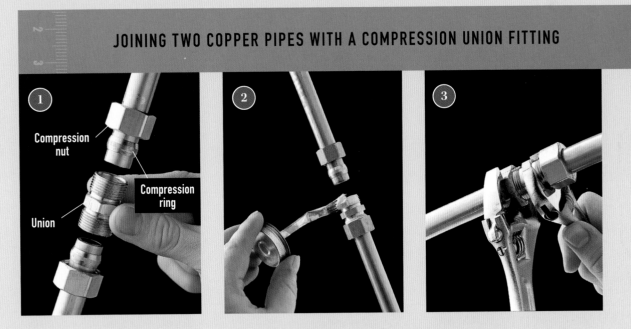

1   Slide compression nuts and rings over the ends of pipes. Place a threaded union between the pipes.

2   Apply a layer of pipe joint compound or Teflon® tape to the union's threads, then screw compression nuts onto the union.

3   Hold the center of the union fitting with an adjustable wrench and use another wrench to tighten each compression nut one complete turn. Turn on the water. If the fitting leaks, tighten the nuts gently.

COMMON TOILET PROBLEMS

A clogged toilet is one of the most common plumbing problems. If a toilet overflows or flushes sluggishly, clear the clog with a plunger or closet auger. If the problem persists, the clog may be in the main waste-vent stack.

Most other toilet problems are fixed easily with minor adjustments that require no disassembly or replacement parts. You can make these adjustments in a few minutes, using simple tools.

If minor adjustments do not fix the problem, further repairs will be needed. The parts of a standard toilet are not difficult to take apart, and most repair projects can be completed in less than an hour.

A recurring puddle of water on the floor around a toilet may be caused by a crack in the toilet base or in the tank. A damaged toilet should be replaced. Installing a new toilet is an easy project that can be finished in three or four hours.

A standard two-piece toilet has an upper tank that is bolted to a base. This type of toilet uses a simple gravity-operated flush system and can easily be repaired using the directions on the following pages. Some one-piece toilets use a complicated, high-pressure flush valve.

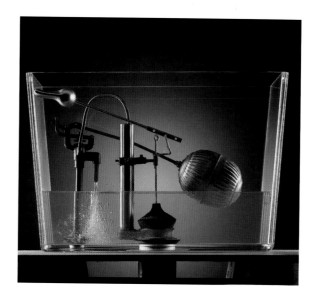

An older toilet may have a tank ball that settles onto the flush valve to stop the flow of water into the bowl. The ball is attached to a lift wire, which is in turn attached to the lift rod. A ballcock valve is usually made of brass, with rubber washers that can wear out. If the ballcock valve malfunctions, you might be able to find old washers to repair it, but replacing both the ballcock and the tank ball with a float-cup assembly and flapper is easier and makes for a more durable repair.

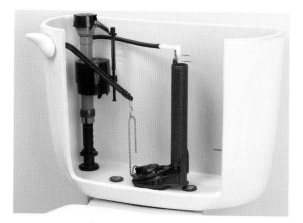

A modern float-cup valve with flapper is inexpensive and made of plastic, but is more reliable than an old ballcock valve and ball.

A pressure-assist toilet has a large vessel that nearly fills the tank. As water enters the vessel, pressure builds up. When the toilet is flushed, this pressure helps push water forcefully down into the bowl. As a result, a pressure-assist toilet provides strong flushing power with minimal water consumption.

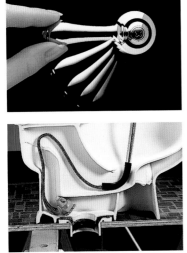

Problems	Repairs
Toilet handle sticks or is hard to push.	1. Adjust lift wires. 2. Clean and adjust handle.
Handle must be held down for entire flush.	1. Adjust handle. 2. Shorten lift chain or wires. 3. Replace waterlogged flapper.
Handle is loose.	1. Adjust handle. 2. Reattach lift chain or lift wires to lever.
Toilet will not flush at all.	1. Make sure water is turned on. 2. Adjust lift chain or lift wires.
Toilet does not flush completely.	1. Adjust lift chain. 2. Adjust water level in tank. 3. Increase pressure on pressure-assisted toilet.
Toilet overflows or flushes sluggishly.	1. Clear clogged toilet. 2. Clear clogged main waste-vent stack.
Toilet runs continuously or there are phantom flushes.	1. Adjust lift wires or lift chain. 2. Replace leaky float ball. 3. Adjust water level in tank. 4. Adjust and clean flush valve. 5. Replace flush valve. 6. Replace flapper. 7. Service pressure-assist valve.
Water on floor around toilet.	1. Tighten tank bolts and water connections. 2. Insulate tank to prevent condensation. 3. Replace wax ring. 4. Replace cracked tank or bowl.
Toilet noisy when filling.	1. Open shutoff valve completely. 2. Replace ballcock and float valve. 3. Refill tube is disconnected.
Weak flush.	1. Clean clogged rim openings. 2. Replace old low-flow toilet.
Toilet rocks.	1. Replace wax ring and bolts. 2. Replace toilet flange.

Making Minor Adjustments

Many common toilet problems can be fixed by making minor adjustments to the handle and the attached lift chain (or lift wires).

If the handle sticks or is hard to push, remove the tank cover and clean the handle-mounting nut. Make sure the lift wires are straight.

If the toilet will not flush completely unless the handle is held down, you may have to remove excess slack in the lift chain.

If the toilet will not flush at all, the lift chain may be broken or may have to be reattached to the handle lever.

A continuously running toilet (page opposite) can be caused by bent lift wires, kinks in a lift chain, or lime buildup on the handle mounting nut. Clean and adjust the handle and the lift wires or chain to fix the problem.

Tools & Materials

Adjustable wrench	Hacksaw
Needlenose pliers	Spray Lubricant
Screwdriver	Small wire brush
Scissors	Vinegar

ADJUSTING A TOILET HANDLE & LIFT CHAIN (OR LIFT WIRES)

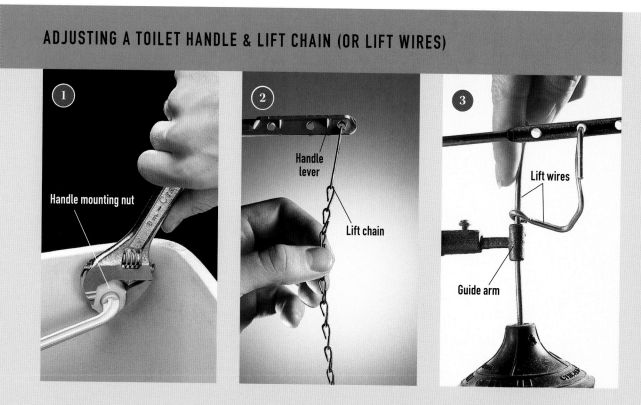

1 — Handle mounting nut

2 — Handle lever, Lift chain

3 — Lift wires, Guide arm

1   Clean and adjust handle-mounting nut so handle operates smoothly. Mounting nut has reversed threads. Loosen nut by turning clockwise; tighten by turning counterclockwise. Remove lime buildup with a brush dipped in vinegar.

2   Adjust lift chain so it hangs straight from handle lever, with about ½" of slack. Remove excess slack in chain by hooking the chain in a different hole in the handle lever or by removing links with needlenose pliers. A broken lift chain must be replaced.

3   Adjust lift wires (found on older toilets without lift chains) so that wires are straight and operate smoothly when handle is pushed. A sticky handle often can be fixed by straightening bent lift wires. You can also buy replacement wires, or replace the whole assembly with a float cup.

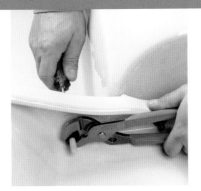

Phantom flushes? Phantom flushes are weak flushes that occur without turning the handle. The flapper may not be completely sealing against the flush valve's seat. Make sure the chain is not tangled, and that the flapper can go all the way down. If that does not solve the problem, shut off water and drain the tank. If the problem persists, the flapper may need to be replaced.

Seat loose? Loose seats are almost always the result of loose nut on the seat bolts. Tighten the nuts with pliers. If the nut is corroded or stripped, replace the bolts and nuts or replace the whole seat.

Seat uncomfortably low? Instead of going to the trouble of raising the toilet or replacing it with a taller model, you can simply replace the seat with a thicker, extended seat.

Bowl not refilling well? The rim holes may be clogged; many toilets have small holes on the underside of the bowl rim, through which water squirts during a flush. If you notice that some of these holes are clogged, use a stiff-bristled brush to clear out debris. You may need to first apply toilet bowl cleaner or mineral cleaner.

Tank fills too slowly? The first place to check is the shutoff valve where the supply tube for the toilet is connected. Make sure it is fully open. If it is, you may need to replace the shutoff—these fittings are fairly cheap and frequently fail to open fully.

Toilet running? Running toilets are usually caused by faulty or misadjusted fill valves, but sometimes the toilet runs because the tank is leaking water into the bowl. To determine if this is happening with your toilet, add a few drops of food coloring to the tank water. If, after a while, the water in the bowl becomes colored, then you have a leak and probably need to replace the rubber gasket at the base of your flush valve.

Reset Tank Water Level

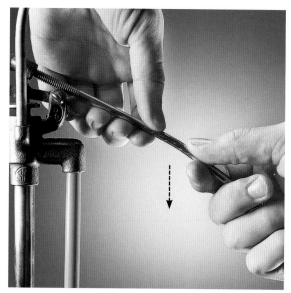

Tank water flowing into the overflow pipe is the sound we hear when a toilet is running. Usually, this is caused by a minor misadjustment that fails to tell the water to shut off when the toilet tank is full. The culprit is a float ball or cup that is adjusted to set a water level in the tank that's higher than the top of the overflow pipe, which serves as a drain for excess tank water. The other photos on this page show how to fix the problem.

A ball float is connected to a float arm that's attached to a plunger on the other end. As the tank fills, the float rises and lifts one end of the float arm. At a certain point, the float arm depresses the plunger and stops the flow of water. By simply bending the float arm downward a bit, you can cause it to depress the plunger at a lower tank water level, solving the problem.

Spring clip

A diaphragm fill valve usually is made of plastic and has a wide bonnet that contains a rubber diaphragm. Turn the adjustment screw clockwise to lower the water level and counterclockwise to raise it.

A float cup fill valve is made of plastic and is easy to adjust. Lower the water level by pinching the spring clip with fingers or pliers and moving the clip and cup down the pull rod and shank. Raise the water level by moving the clip and cup upward.

What If the Flush Stops Too Soon?

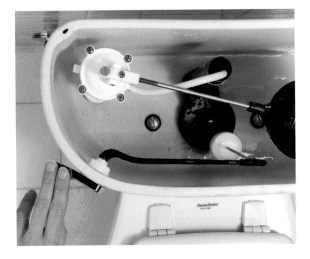

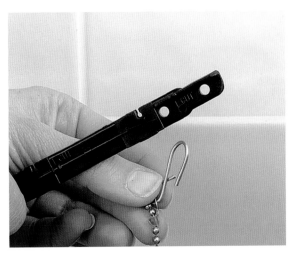

Sometimes there is plenty of water in the tank, but not enough of it makes it to the bowl before the flush valve shuts off the water from the tank. Modern toilets are designed to leave some water in the tank, since the first water that leaves the tank does so with the most force. (It's pressed out by the weight of the water on top.) To increase the duration of the flush, shorten the length of the chain between the flapper and the float (yellow in the model shown).

The handle lever should pull straight up on the flapper. If it doesn't, reposition the chain hook on the handle lever. When the flapper is covering the opening, there should be just a little slack in the chain. If there is too much slack, shorten the chain and cut off excess with the cutters on your pliers.

If the toilet is not completing flushes and the lever and chain for the flapper or tank ball are correctly adjusted, the problem could be that the handle mechanism needs cleaning or replacement. Remove the chain/linkage from the handle lever. Remove the nut on the backside of the handle with an adjustable wrench. It unthreads clockwise (the reverse of standard nuts). Remove the old handle from the tank.

Unless the handle parts are visibly broken, try cleaning them with an old toothbrush dipped in white vinegar. Replace the handle and test the action. If it sticks or is hard to operate, replace it. Most replacement handles come with detailed instructions that tell you how to install and adjust them.

1   Toilet fill valves wear out eventually. They can be repaired, but it's easier and a better fix to just replace them. Before removing the old fill valve, shut off the water supply at the fixture stop valve located on the tube that supplies water to the tank. Flush the toilet and sponge out the remaining water. Loosen the nut and disconnect the supply tube, then loosen and remove the mounting nut.

2   If the fill valve spins while you turn the mounting nut, you may need to hold it still with locking pliers. Lift out the fill valve. In the case of an old ballcock valve, the float ball will likely come out as well. When replacing an old valve like this, you will likely also need to replace the flush valve (see pages 68-69).

3   The new fill valve must be installed so the critical level ("CL") mark is at least 1" above the overflow pipe (see inset). Slip the shank washer on the threaded shank of the new fill valve and place the valve in the hole so the washer is flat on the tank bottom. Compare the locations of the "CL" mark and the overflow pipe.

4   Adjust the height of the fill valve shank so the "CL" line and overflow pipe will be correctly related. Different products are adjusted in different ways—the fill valve shown here telescopes when it's twisted.

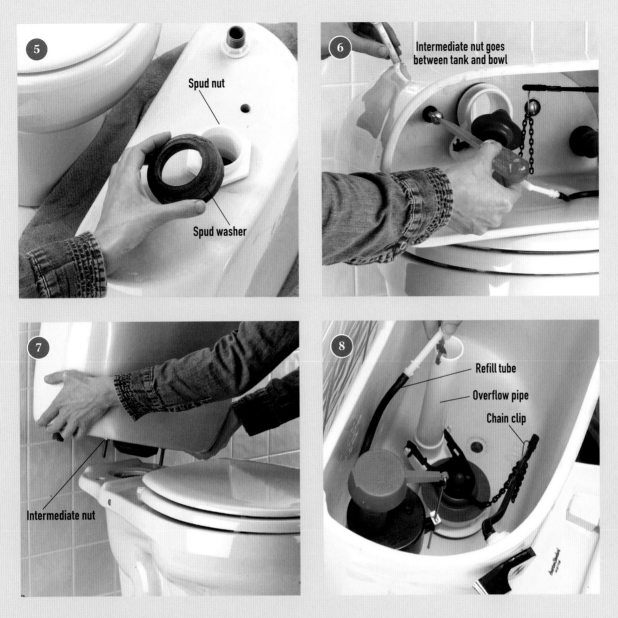

5 Spud nut

Spud washer

6 Intermediate nut goes between tank and bowl

7 Intermediate nut

8 Refill tube

Overflow pipe

Chain clip

5   Position the flush valve flapper below the handle lever arm and secure it to the tank from beneath with the spud nut. Tighten the nut one-half turn past hand tight with a spud wrench or large channel-type pliers. Overtightening may cause the tank to break. Put the new spud washer over the spud nut, small side down.

6   With the tank lying on its back, thread a rubber washer onto each tank bolt and insert it into the bolt holes from inside the tank. Then, thread a brass washer and hex nut onto the tank bolts from below and tighten them to a quarter turn past hand tight. Do not overtighten.

7   With the hex nuts tightened against the tank bottom, carefully lower the tank over the bowl and set it down so the spud washer seats neatly over the water inlet in the bowl and the tank bolts fit through the holes in the bowl flange. Secure the tank to the bowl with a rubber washer, brass washer, and nut or wing nut at each bolt end. Press the tank to level as you hand-tighten the nuts. Hook up the water supply at the fill valve inlet.

8   Connect the chain clip to the handle lever arm and adjust the number of links to allow for a little slack in the chain when the flapper is closed. Leave a little tail on the chain for adjusting, cutting off remaining excess. Attach the refill tube to the top of the overflow pipe the same way it had been attached to the previous refill pipe. Turn on the water supply at the stop valve and test the flush. (Some flush valve flappers are adjustable.)

1   Plunging is the easiest way to remove "natural" blockages. Take time to lay towels around the base of the toilet and remove other objects to a safe, dry location, since plunging may result in splashing. Often, allowing a very full toilet to sit for twenty or thirty minutes will permit some of the water to drain to a less precarious level.

2   There should be enough water in the bowl to completely cover the plunger. Fold out the skirt from inside the plunger to form a better seal with the opening at the base of the bowl. Pump the plunger vigorously half-a-dozen times, take a rest, and then repeat. Try this for four to five cycles.

3   If you force enough water out of the bowl that you are unable to create suction with the plunger, put a controlled amount of water in the bowl by lifting up on the flush valve in the tank. Resume plunging. When you think the drain is clear, you can try a controlled flush, with your hand ready to close the flush valve should the water threaten to spill out of the bowl. Once the blockage has cleared, dump a five-gallon pail of water into the toilet to blast away any residual debris.

Force Cups

A flanged plunger (force cup) fits into the mouth of the toilet trap and creates a tight seal so you can build up enough pressure in front of the plunger to dislodge the blockage and send it on its way.

Protective rubber boot

1   Place the business end of the auger firmly in the bottom of the toilet bowl with the auger tip fully withdrawn. A rubber sleeve will protect the porcelain at the bottom bend of the auger. The tip will be facing back and up, which is the direction the toilet trap takes.

2   Rotate the handle on the auger housing clockwise as you push down on the rod, advancing the rotating auger tip up into the back part of the trap. You may work the cable backward and forward as needed, but keep the rubber boot of the auger firmly in place in the bowl. When you feel resistance, indicating you've snagged the object, continue rotating the auger counterclockwise as you withdraw the cable and the object.

3   Fully retract the auger until you have recovered the object. This can be frustrating at times, but it is still a much easier task than the alternative—to remove the toilet and go fishing.

Closet Augers

A closet auger is a semirigid cable housed in a tube. The tube has a bend at the end so it can be snaked through a toilet trap (without scratching it) to snag blockages.

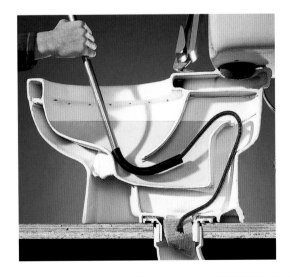

Kitchen Faucets

Tools & Materials

Adjustable wrench
Basin wrench or
    channel-type pliers
Hacksaw
Faucet
Putty knife
Screwdriver
Silicone caulk
Scouring pad
Scouring cleaner
Plumber's putty
Flexible vinyl or braided steel
    supply tubes
Drain components
Penetrating oil

Modern kitchen faucets tend to be single-handle models, often with useful features such as a pull-out head that functions as a sprayer. This Pfister model comes with an optional mounting plate that conceals sink holes when mounted on a predrilled sink flange.

MOST NEW KITCHEN FAUCETS feature single-handle control levers and washerless designs that rarely require maintenance. Additional features include brushed metallic finishes, detachable spray nozzles, or even push-button controls.

Connect the faucet to hot and cold water lines with easy-to-install flexible supply tubes made from vinyl or braided steel. If your faucet has a separate sprayer, install the sprayer first. Pull the sprayer hose through the sink opening and attach it to the faucet body before installing the faucet.

Where local codes allow, use plastic tubes for drain hookups. A wide selection of extensions and angle fittings lets you easily plumb any sink configuration. Manufacturers offer kits that contain all the fittings needed for attaching a food disposer or dishwasher to the sink drain system.

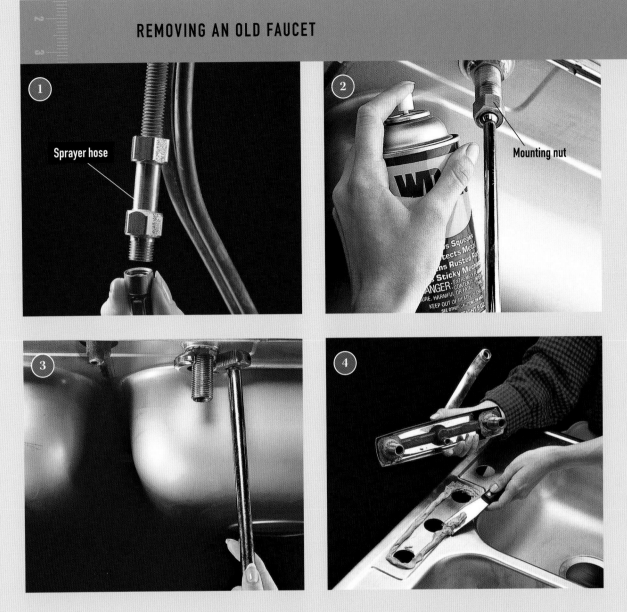

1   To remove the old faucet, start by clearing out the cabinet under the sink and laying down towels. Turn off the hot and cold stop valves and open the faucet to make sure the water is off. Detach the sprayer hose from the faucet sprayer nipple and unscrew the retaining nut that secures the sprayer base to the sink deck. Pull the sprayer hose out through the sink deck opening.

2   Spray the mounting nuts that hold the faucet or faucet handles (on the underside of the sink deck) with penetrating oil for easier removal. Let the oil soak in for a few minutes. If the nut is rusted and stubbornly stuck, you may need to drill a hole in its side, then tap the hole with a hammer and screwdriver to loosen it.

3   Unhook the supply tubes at the stop valves. Don't reuse old chrome supply tubes. If the stops are missing or unworkable, replace them. Then remove the coupling nuts and the mounting nuts on the tailpieces of the faucet with a basin wrench or channel-type pliers.

4   Pull the faucet body from the sink. Remove the sprayer base if you wish to replace it. Scrape off old putty or caulk with a putty knife and clean off the sink with a scouring pad and an acidic scouring cleaner like Bar Keepers Friend. Tip: Scour stainless steel with a back-and-forth motion to avoid leaving unsightly circular markings.

INSTALLING A PULLOUT KITCHEN SINK FAUCET

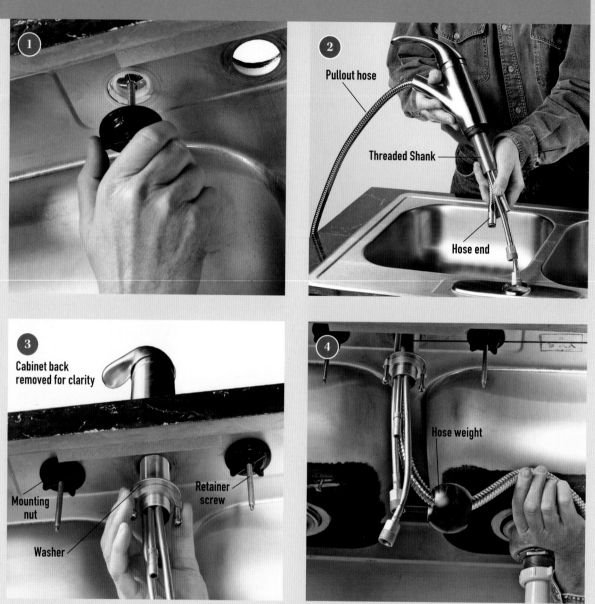

1   Install the base plate (if your faucet has one) onto the sink flange so it is centered. Have a helper hold it straight from above as you tighten the mounting nuts that secure the base plate from below. Make sure the plastic gasket is centered under the base plate. These nuts can be adequately tightened by hand.

2   Retract the pullout hose by drawing it out through the faucet body until the fitting at the end of the hose is flush with the bottom of the threaded faucet shank. Insert the shank and the supply tubes down through the top of the deck plate.

3   Slip the mounting nut and washer over the free ends of the supply tubes and pullout hose, then thread the nut onto the threaded faucet shank. Hand tighten. Tighten the retainer screws with a screwdriver to secure the faucet.

4   Slide the hose weight onto the pullout hose (the weight helps keep the hose from tangling and it makes it easier to retract).

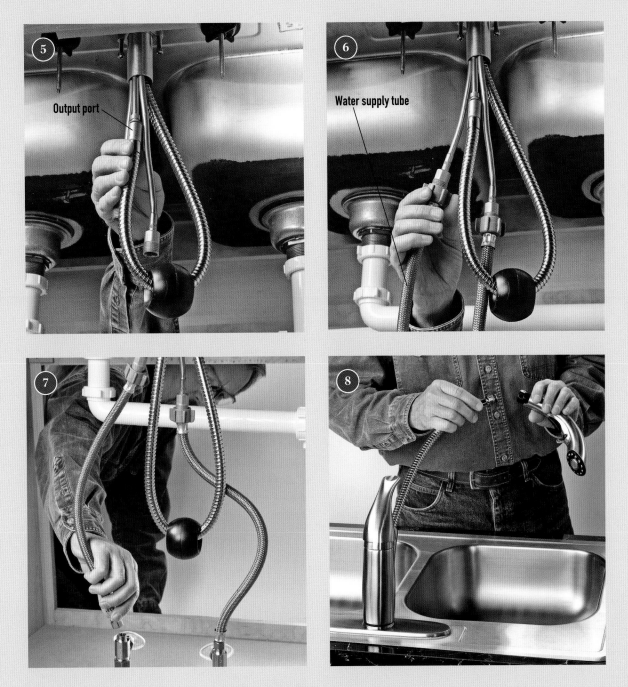

5   Connect the end of the pullout hose to the outlet port on the faucet body using a quick connector fitting.

6   Hook up the water supply tubes to the faucet inlets. Make sure the tubes are long enough to reach the supply risers without stretching or kinking.

7   Connect the supply tubes to the supply risers at the stop valves. Make sure to get the hot lines and cold lines attached correctly.

8   Attach the spray head to the end of the pullout hose and turn the fitting to secure the connection. Turn on water supply and test. Tip: Remove the aerator in the tip of the spray head and run hot and cold water to flush out any debris.

Identifying Your Faucet and the Parts You Need

A leaky faucet is the most common home plumbing problem. Fortunately, repair parts are available for almost every type of faucet, from the oldest to the newest, and installing these parts is usually easy. But if you don't know your make and model, the hardest part of fixing a leak may be identifying your faucet and finding the right parts. Don't make the common mistake of thinking that any similar-looking parts will do the job; you've got to get exact replacements.

There are so many faucet types that even experts have trouble classifying them into neat categories. Two-handle faucets are either compression (stem) or washerless two-handle. Single-handle faucets are classified as mixing cartridge; ball; disc; or disc/cartridge.

A single-handle faucet with a rounded, dome-shaped cap is often a ball type. If a single-handle faucet has a flat top, it is likely a cartridge or a ceramic disc type. An older two-handle faucet is likely of the compression type; newer two-handle models use washerless cartridges. Shut off the water, and test to verify that the water is off. Dismantle the faucet carefully. Look for a brand name: it may be clearly visible on the baseplate, or may be printed on an inner part, or it may not be printed anywhere. Put all the parts into a reliable plastic bag and take them to your home center or plumbing supply store. A knowledgeable salesperson can help you identify the parts you need.

If you cannot find what you are looking for at a local store, check online faucet sites or the manufacturers' sites; they often have step-by-step instructions for identifying what you need. Note that manufacturers' terminology may not match the terms we use here. For example, the word "cartridge" may refer to a ceramic-disc unit.

Most faucets have repair kits, which include all the parts you need, and sometimes a small tool as well. Even if some of the parts in your faucet look fine, it's a good idea to install the parts provided by the kit, to ensure against future wear.

Repair Tips

If water flow is weak, unscrew the aerator at the tip of the spout. If there is sediment, then dirty water is entering the faucet, which could damage the faucet's inner workings.

To remove handles and spouts, work carefully and look for small screw heads. You often need to first pry off a cap on top, but not always. Parts may be held in place with small setscrews.

Cleaning and removing debris can sometimes solve the problem of low water flow, and occasionally can solve a leak as well.

Apply plumber's grease (also known as faucet grease or valve grease), to new parts before installing them. Be especially sure to coat rubber parts like O-rings and washers.

Compression Faucets

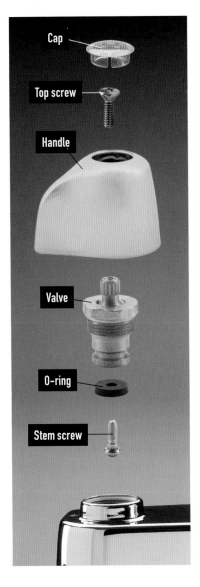

Cap

Top screw

Handle

Valve

O-ring

Stem screw

Pry off the cap on top of the handle and remove the screw that holds the cap onto the stem. Pull the handle up and out. Use an adjustable wrench or pliers to unscrew the stem and pull it out.

If the handle is stuck, try applying mineral cleaner from above. If that doesn't work, you may need to buy a handle puller. With the cap and the hold-down screw removed, position the wings of the puller under the handle and tighten the puller to slowly pull the handle up.

A compression faucet has a stem assembly that includes a retaining nut, threaded spindle, O-ring, stem washer, and stem screw. Dripping at the spout occurs when the washer becomes worn. Leaks around the handle are caused by a worn O-ring.

If washers wear out quickly, the seat is likely worn. Use a seat wrench to unscrew the seat from inside the faucet. Replace it with an exact duplicate. If replacing the washer and O-ring doesn't solve the problem, you may need to replace the entire stem.

Remove the screw that holds the rubber washer in place, and pry out the washer. Replace a worn washer with an exact replacement—one that is the same diameter, thickness, and shape.

Washerless Two-handle Faucet

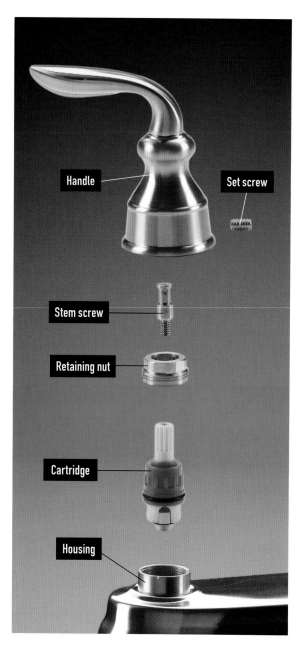

Handle

Set screw

Stem screw

Retaining nut

Cartridge

Housing

Retaining nut

Remove the faucet handle and withdraw the old cartridge. Make a note of how the cartridge is oriented before you remove it. Purchase a replacement cartridge.

Almost all two-handle faucets made today are "washerless." Instead of an older-type compression stem, there is a cartridge, usually with a plastic casing. Many of these cartridges contain ceramic discs, while others have metal or plastic pathways. No matter the type of cartridge, the repair is the same; instead of replacing small parts, you simply replace the entire cartridge.

Install the replacement cartridge. Clean the valve seat first and coat the valve seat and O-rings with faucet grease. Be sure the new cartridge is in the correct position, with its tabs seated in the slotted body of the faucet. Re-assemble the valve and handles.

One-Handle Cartridge Faucets

Cap

Handle

Cap screw

Retaining nut

Spout

O-ring and gasket

Retainer clip

Cartridge

O-rings

Faucet body

To remove the spout, pry off the handle's cap and remove the screw below it. Pull the handle up and off. Use a crescent wrench to remove the pivot nut.

Lift out the spout. If the faucet has a diverter valve, remove it as well. Use a screwdriver to pry out the retainer clip, which holds the cartridge in place.

Remove the cartridge. If you simply pull up with pliers, you may leave part of the stem in the faucet body. If that happens, replace the cartridge and buy a stem puller made for your model.

Gently pry out and replace all O-rings on the faucet body. Smear plumber's grease onto the new replacement cartridge and the new O-rings, and reassemble the faucet.

Single-handle cartridge faucets like this work by moving the cartridge up and down and side to side, which opens up pathways to direct varying amounts of hot and cold water to the spout. Moen, Price-Pfister, Delta, Peerless, Kohler, and others make many types of cartridges, some of which look very different from this one.

Repair Tips

Here is one of many other types of single-handle cartridges. In this model, all the parts are plastic except for the stem, and it's important to note the direction in which the cartridge is aligned. If you test the faucet and the hot and cold are reversed, disassemble and realign the cartridge.

Ball Faucets

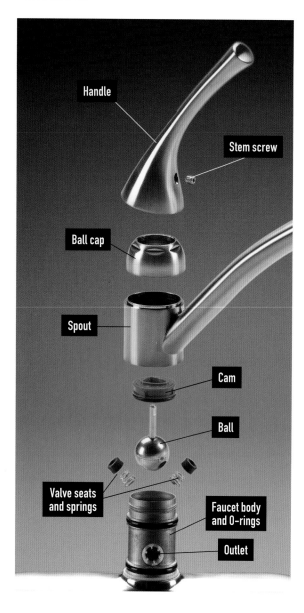

Remove the old ball and cam after removing the faucet handle and ball cap. Some faucets may require a ball faucet tool to remove the handle. Otherwise, simply use a pair of channel-type pliers to twist off the ball cap.

The ball-type faucet is used by Delta, Peerless, and a few others. The ball fits into the faucet body and is constructed with three holes (not visible here)—a hot inlet, a cold inlet, and the outlet, which fills the valve body with water that then flows to the spout or sprayer. Depending on the position of the ball, each inlet hole is open, closed, or somewhere in-between. The inlet holes are sealed to the ball with valve seats, which are pressed tight against the ball with springs. If water drips from the spout, replace the seats and springs. Or go ahead and purchase an entire replacement kit and replace all or most of the working parts.

Pry out the neoprene valve seals and springs. Place thick towels around the faucet. Slowly turn on the water to flush out any debris in the faucet body. Replace the seals and springs with new parts. Also replace the O-rings on the valve body. You may want to replace the ball and cam, too, especially if you're purchasing a repair kit. Coat all rubber parts in faucet grease, and reassemble the faucet.

Disc Faucets

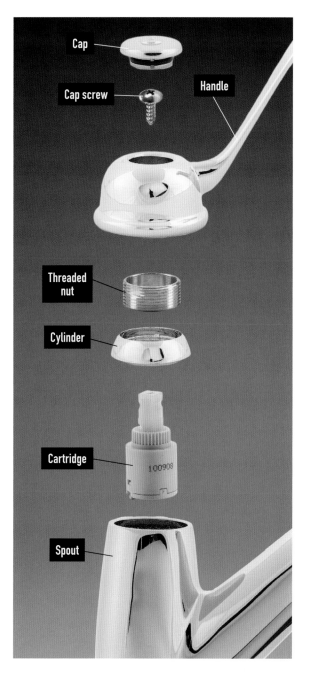

Replace the cylinder with a new one, coating the rubber parts with faucet grease before installing the new cylinder. Make sure the rubber seals fit correctly in the cylinder openings before you install the cylinder. Assemble the faucet handle.

Other Cartridges

Many modern cartridges do not have seals or O-rings that can be replaced, and some have a ball rather than a ceramic disk inside. For the repair, the cartridge's innards do not matter; just replace the whole cartridge.

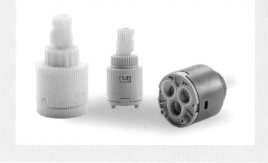

Disc-type faucets are the most common single-handle faucets currently being made. A pair of ceramic discs encased in a cylinder often referred to as a "cartridge" rub together as they rotate to open ports for hot and cold water. The ceramic discs do wear out in time, causing leaks, and there is only one solution—replace the disc unit (or cartridge). This makes for an easy—through comparatively expensive—repair.

Bathroom Faucets

Bathroom sink faucets come in two basic styles: the widespread with independent handles and spout (top); and the single-body, deck-mounted version (left).

ONE-PIECE FAUCETS, WITH EITHER ONE or two handles, are the most popular fixtures for bathroom installations.

"Widespread" faucets with separate spout and handles are being installed with increasing frequency, however. Because the handles are connected to the spout with flex tubes that can be 18" or longer, widespread faucets can be arranged in many ways.

Bathroom Faucet & Drain Hookups

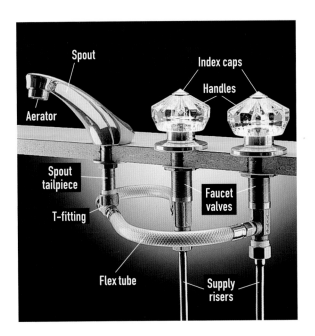

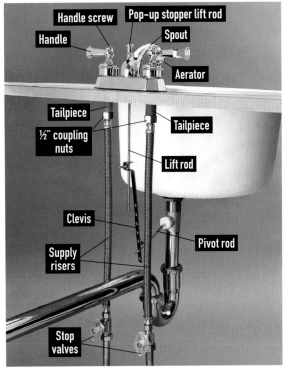

Widespread lavatory faucets have valves that are independent from the spout so they can be configured however you choose, provided that your flex tube connectors are long enough to span the distance.

Single-body lavatory faucets have both valves and the spout permanently affixed to the faucet body. They do not offer flexibility in configurations, but they are very simple to install.

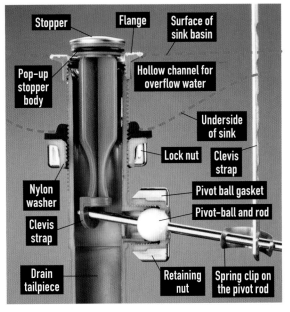

The pop-up stopper fits into the drain opening so the stopper will close tightly against the drain flange when the pop-up handle is lifted up.

The linkage that connects the pop-up stopper to the pop-up handle fits into a male-threaded port in the drain tailpiece. Occasionally the linkage will require adjustment or replacement.

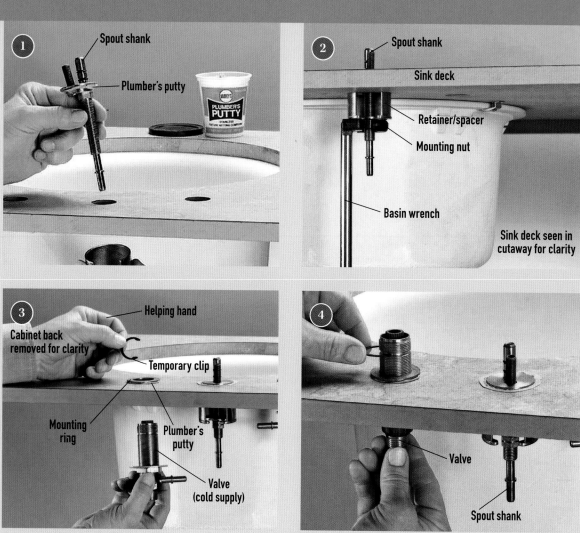

1 Insert the shank of the faucet spout through one of the holes in the sink deck (usually the center hole, but you can offset it in one of the end holes if you prefer). If the faucet is not equipped with seals or O-rings for the spout and handles, pack plumber's putty on the undersides before inserting the valves into the deck. If you are installing the widespread faucet in a new sink deck, drill three holes of the size suggested by the faucet manufacturer.

2 In addition to mounting nuts, many spout valves for widespread faucets have an open-retainer fitting that goes between the underside of the deck and the mounting nut. Others have only a mounting nut. In either case, tighten the mounting nut with pliers or a basin wrench to secure the spout valve. You may need a helper to keep the spout centered and facing forward.

3 Mount the valves to the deck using whichever method the manufacturer specifies (it varies quite a bit). In the model seen here, a mounting ring is positioned over the deck hole (with plumber's putty seal) and the valve is inserted from below. A clip snaps onto the valve from above to hold it in place temporarily (you'll want a helper for this).

4 From below, thread the mounting nuts that secure the valves to the sink deck. Make sure the cold water valve (usually has a blue cartridge inside) is in the right-side hole (from the front) and the hot water valve (red cartridge) is in the left hole. Install both valves.

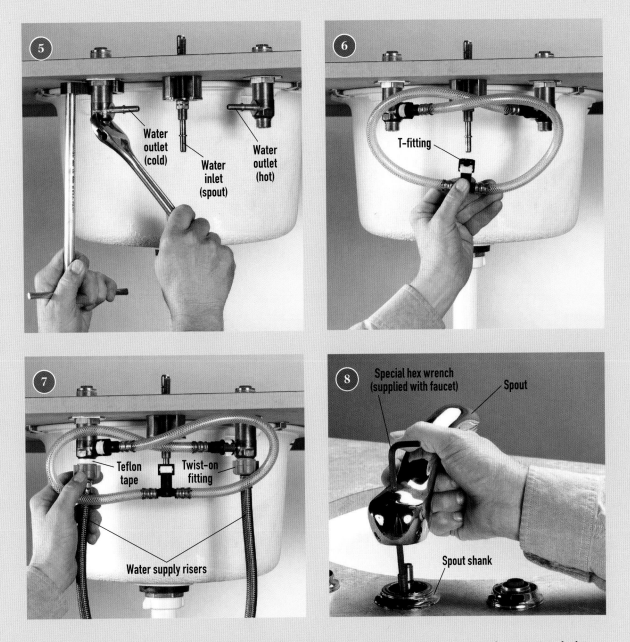

5 Once you've started the nut on the threaded valve shank, secure the valve with a basin wrench, squeezing the lugs where the valve fits against the deck. Use an adjustable wrench to finish tightening the lock nut onto the valve. The valves should be oriented so the water outlets are aimed at the inlet on the spout shank.

6 Attach the flexible supply tubes (supplied with the faucet) to the water outlets on the valves. Some twist onto the outlets, but others (like the ones above) click into place. The supply hoses meet in a T-fitting that is attached to the water inlet on the spout.

7 Attach flexible braided-metal supply risers to the water stop valves and then attach the tubes to the inlet port on each valve (usually with Teflon tape and a twist-on fitting at the valve end of the supply riser).

8 Attach the spout. The model shown here comes with a special hex wrench that is threaded through the hole in the spout where the lift rod for the pop-up drain will be located. Once the spout is seated cleanly on the spout shank, you tighten the hex wrench to secure the spout. Different faucets will use other methods to secure the spout to the shank.

Continued

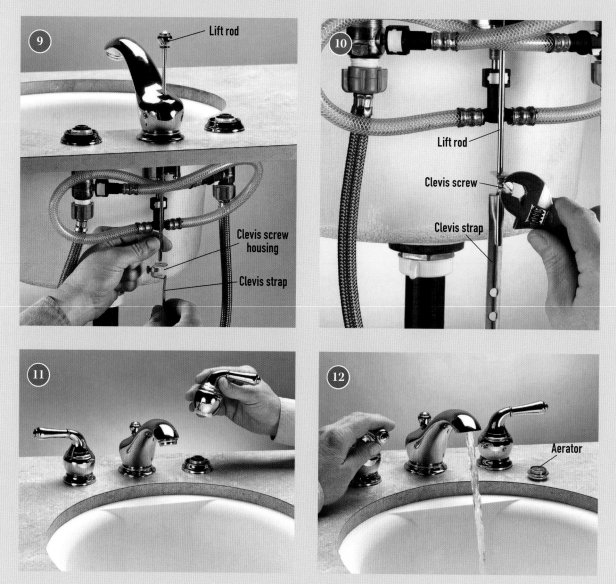

9   If your sink did not have a pop-up stopper, you'll need to replace the sink drain tailpiece with a pop-up stopper body (often supplied with the faucet). Insert the lift rod through the hole in the back of the spout and, from below, thread the pivot rod through the housing for the clevis screw.

10 Attach the clevis strap to the pivot rod that enters the pop-up drain body, and adjust the position of the strap so it raises and lowers properly when the lift rod is pulled up. Tighten the clevis screw at this point. It's hard to fit a screwdriver in here, so you may need to use a wrench or pliers.

11 Attach the faucet handles to the valves using whichever method is required by the faucet manufacturer. Most faucets are designed with registration methods to ensure that the handles are symmetrical and oriented in an ergonomic way once you secure them to the valves.

12 Turn on the water supply and test the faucet. Remove the faucet aerator and run the water for 10 to 20 seconds so any debris in the lines can clear the spout. Replace the aerator.

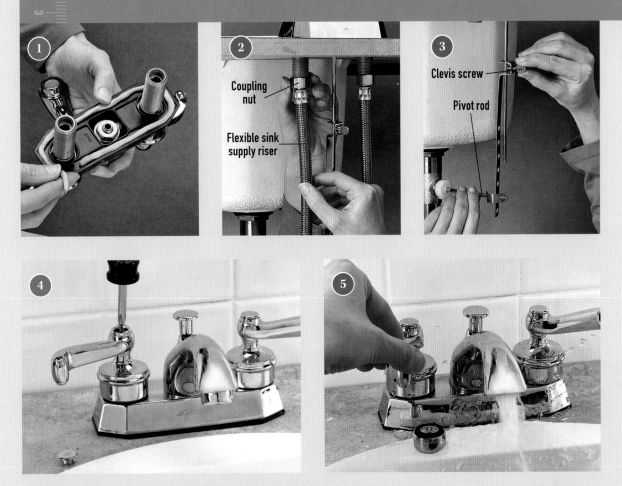

1   High-quality faucets come with flexible plastic gaskets that create a durable watertight seal at the bottom of the faucet, where it meets the sink deck. However, an inexpensive faucet may have a flimsy-looking foam seal that doesn't do a good job of sealing and disintegrates after a few years. If that is the case with your faucet, discard the seal and press a ring of plumber's putty into the sealant groove on the underside of the faucet body.

2   Insert the faucet tailpieces through the holes in the sink. From below, thread washers and mounting nuts over the tailpieces, then tighten the mounting nuts with a basin wrench until snug. Put a dab of pipe joint compound on the threads of the stop valves and thread the metal nuts of the flexible supply risers to these. Wrench tighten about a half-turn past hand tight. Overtightening these nuts will strip the threads. Now tighten the coupling nuts to the faucet tailpieces with a basin wrench.

3   Slide the lift rod of the new faucet into its hole behind the spout. Thread it into the clevis past the clevis screw. Push the pivot rod all the way down so the stopper is open. With the lift rod also all the way down, tighten the clevis to the lift rod.

4   Grease the fluted valve stems with faucet grease, then put the handles in place. Tighten the handle screws firmly, so they won't come loose during operation. Cover each handle screw with the appropriate index cap—Hot or Cold.

5   Unscrew the aerator from the end of the spout. Turn the hot and cold water taps on full. Turn the water back on at the stop valves and flush out the faucet for a couple of minutes before turning off the water at the faucet. Check the riser connections for drips. Tighten a compression nut only until the drip stops. Replace the aerator.

INSTALLING A POP-UP DRAIN

1 — Lock nuts · Pop-up drain tailpiece · Trap arm · Trap J-bend

2 — Bottom of sink · Clevis · Spring clip · Cap · Stopper body · Ball-and-pivot rod

3 — Stopper · Flange

4 — Wrap tape in clockwise direction · Stopper body

1   Put a basin under the trap to catch water. Loosen the nuts at the outlet and inlet to the trap J-bend by hand or with channel-type pliers and remove the bend. The trap will slide off the pop-up body tailpiece when the nuts are loose. Keep track of washers and nuts and their up/down orientation by leaving them on the tubes.

2   Unscrew the cap holding the ball-and-pivot rod in the pop-up body and withdraw the ball. Compress the spring clip on the clevis and withdraw the pivot rod from the clevis.

3   Remove the pop-up stopper. Then, from below, remove the lock nut on the stopper body. If needed, keep the flange from turning by inserting a large screwdriver in the drain from the top. Thrust the stopper body up through the hole to free the flange from the basin, and then remove the flange and the stopper body.

4   Clean the drain opening above and below, and then thread the locknut all the way down the new pop-up body, followed by the flat washer and the rubber gasket (beveled side up). Wrap three layers of Teflon tape clockwise onto the top of the threaded body. Make a ½"-dia. snake from plumber's putty, form it into a ring, and stick the ring underneath the drain flange.

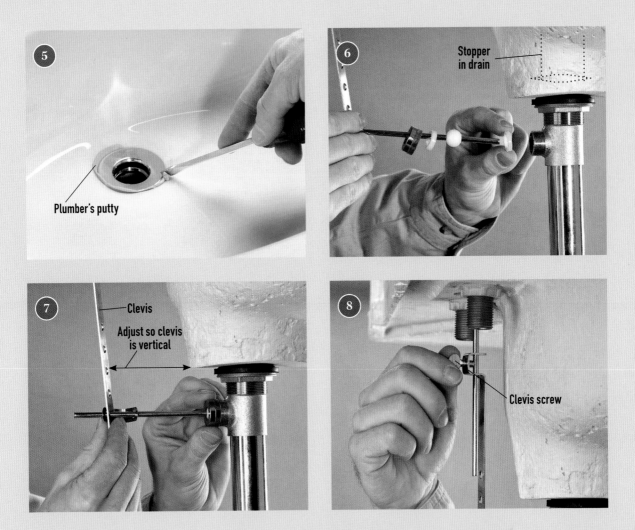

5 From below, face the pivot rod opening directly back toward the middle of the faucet and pull the body straight down to seat the flange. Thread the locknut/washer assembly up under the sink, then fully tighten the locknut with channel-type pliers. Do not twist the flange in the process, as this can break the putty seal. Clean off the squeezeout of plumber's putty from around the flange.

6 Drop the pop-up stopper into the drain hole so the hole at the bottom of its post is closest to the back of the sink. Put the beveled nylon washer into the opening in the back of the pop-up body with the bevel facing back.

7 Put the cap behind the ball on the pivot rod as shown. Sandwich a hole in the clevis with the spring clip and thread the long end of the pivot rod through the clip and clevis. Put the ball end of the pivot rod into the pop-up body opening and into the hole in the stopper stem. Screw the cap on to the pop-up body over the ball.

8 Loosen the clevis screw holding the clevis to the lift rod. Push the pivot rod all the way down (which fully opens the pop-up stopper). With the lift rod also all the way down, tighten the clevis screw to the rod. If the clevis runs into the top of the trap, cut it short with your hacksaw or tin snips. Reassemble the J-bend trap.

Always Test Drain for Leaks

To make sure the sink will not leak, do a thorough test. Close the stopper and turn on the faucet to fill the bowl. Once full, open the stopper and look carefully beneath the sink. Feel the trap parts; they should be dry. If there is any indication of moisture, tighten trap parts as needed.

Kitchen Drains & Traps

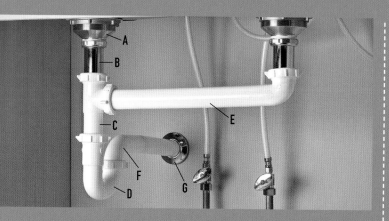

Tools & Materials

Flat screwdriver
Spud wrench
Trap arm
Mineral spirits
Cloth
Strainer kit
Plumber's putty
Teflon tape
Washers
Waste-T fitting
P-trap
Saw
Miter box

Kitchen sink drains include a strainer body (A), tailpiece (B), waste-T (C), P-trap (D), outlet drain line (E), trap arm (F), and wall stubout with coupling (G).

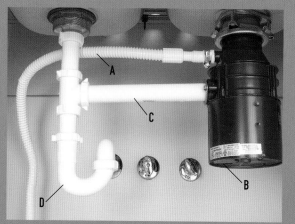

In this arrangement, the dishwasher drain hose (A) attaches to the food disposer (B), and a trap arm (C) leads from the disposer to the P-trap (D).

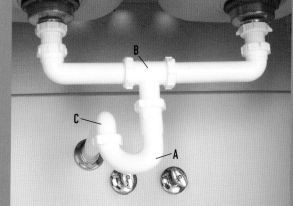

A "center tee" arrangement has a single P-trap (A) that is connected to a waste T (B) and the trap arm (C).

KITCHEN TRAPS, ALSO CALLED sink drains or trap assemblies, are made of 1½-inch pipes (also called tubes), slip washers, and nuts, so they can be easily assembled and disassembled. Most plastic types can be tightened by hand, with no wrench required. Pipes made of chromed brass will corrode in time, and rubber washers will crumble, meaning they need to be replaced. Plastic pipes and plastic washers last virtually forever. All traps are liable to get bumped out of alignment; when

this happens, they should be taken apart and reassembled.

A trap's configuration depends on how many bowls the sink has, whether or not you have a food disposer and/or a dishwasher drain line, and local codes. On this page we show three of the most common assembly types. T fittings on these traps often have a baffle, which reduces the water flow somewhat. Check local codes to make sure your trap is compliant.

Tip: Drain Kits

Kits for installing a new sink drain include all the pipes, slip fittings, and washers you'll need to get from the sink tailpieces (most kits are equipped for a double bowl kitchen sink) to the trap arm that enters the wall or floor. For wall trap arms, you'll need a kit with a P-trap. Both drains normally are plumbed to share a trap. Chromed brass or PVC with slip fittings let you adjust the drain more easily and pull it apart and then reassemble if there is a clog. Some pipes have fittings on their ends that eliminate the need for a washer. Kitchen sink drains and traps should be 1½" o.d. pipe—the 1¼" pipe is for lavatories and doesn't have enough capacity for a kitchen sink.

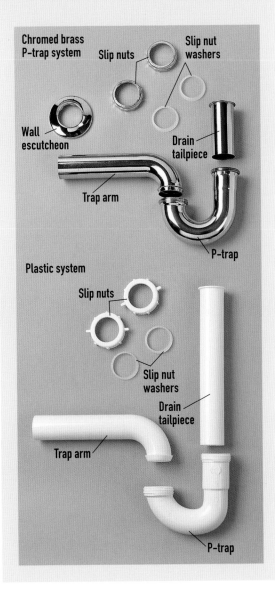

Chromed brass P-trap system

Slip nuts

Slip nut washers

Wall escutcheon

Drain tailpiece

Trap arm

P-trap

Plastic system

Slip nuts

Slip nut washers

Drain tailpiece

Trap arm

P-trap

Tips for Choosing Drains

To make sure the sink will not leak, do a thorough test. Close the stopper and turn on the faucet to fill the bowl. Once full, open the stopper and look carefully beneath the sink. Feel the trap parts; they should be dry. If there is any indication of moisture, tighten trap parts as needed.

Chromed brass

Heavy plastic

Light-duty plastic

Wall thickness varies in sink drain pipes. The thinner plastic material is cheaper and more difficult to obtain a good seal than with the thicker, more expensive tubing. The thin product is best reserved for lavatory drains, which are far less demanding.

Slip joints are formed by tightening a male-threaded slip nut over a female-threaded fitting, trapping and compressing a beveled nylon washer to seal the joint.

Use a spud wrench to tighten the strainer body against the underside of the sink bowl. Normally, the strainer flange has a layer of plumber's putty to seal beneath it above the sink drain, and a pair of washers (one rubber, one fibrous) to seal below.

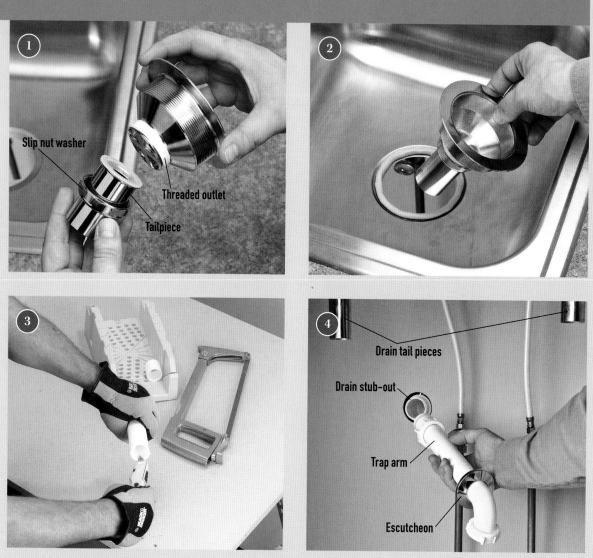

1   If you are replacing the sink strainer body, remove the old one and clean the top and bottom of the sink deck around the drain opening with mineral spirits. Attach the drain tailpiece to the threaded outlet of the strainer body, inserting a nonbeveled washer between the parts if your strainer kits include one. Lubricate the threads or apply Teflon tape so you can get a good, snug fit.

2   Apply plumber's putty around the perimeter of the drain opening and seat the strainer assembly into it. Add washers below as directed and tighten the strainer locknut with a spud wrench (see photo, previous page) or by striking the mounting nubs at the top of the body with a flat screwdriver.

3   You may need to cut a trap arm or drain tailpiece to length. Cut metal tubing with a hacksaw. Cut plastic tubing with a handsaw, power miter saw, or a hand miter box and a backsaw or hacksaw. You can use a tubing cutter for any material. Deburr the cut end of plastic tubing with a utility knife.

4   Attach the trap arm to the male-threaded drain stubout in the wall, using a slip nut and beveled compression washer. The outlet for the trap arm should point downward. Note: The trap arm must be lower on the wall than any of the horizontal lines in the set-up, including lines to dishwasher, disposer, or the outlet line to the second sink bowl.

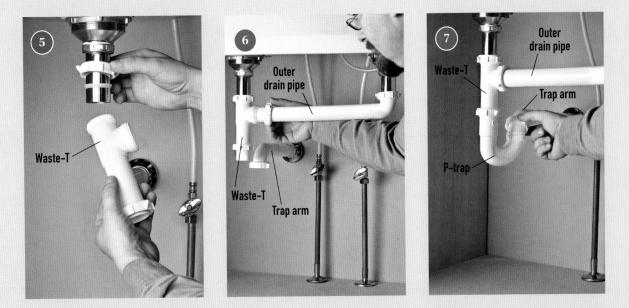

5   Attach a waste-T-fitting to the drain tailpiece, orienting the opening in the fitting side so it will accept the outlet drain line from the other sink bowl. If the waste-T is higher than the top of the trap arm, remove it and trim the drain tailpiece.

6   Join the short end of the outlet drain pipe to the tailpiece for the other sink bowl and then attach the end of the long run to the opening in the waste-T. The outlet tube should extend into the T ½"—make sure it does not extend in far enough to block water flow from above.

7   Attach the long leg of a P-trap to the waste-T and attach the shorter leg to the downward-facing opening of the trap arm. Adjust as necessary and test all joints to make sure they are still tight, and then test the system.

Variation: Drain in Floor

If your drain stubout comes up out of the floor instead of the wall, you have an S-trap instead of a P-trap. This arrangement is illegal in many parts of the country, because a heavy surge of water can siphon the trap dry, rendering it unable to trap gases. However, if after draining the sink you run a slow to moderate stream of water for a few seconds, the trap will fill. An S-trap has two trap pipes that lead to a straight vertical pipe.

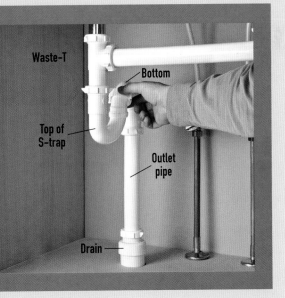

CLOGGED SINK DRAINS

Every sink has a drain trap and a fixture drain line. Sink clogs usually are caused by a buildup of soap and hair in the trap or fixture drain line. Remove clogs by using a plunger, disconnecting and cleaning the trap or using a hand auger.

Many sinks hold water with a mechanical plug called a pop-up stopper. If the sink will not hold standing water, or if water in the sink drains too slowly, the pop-up stopper must be cleaned and adjusted.

Clogged lavatory sinks can be cleared with a plunger (not to be confused with a flanged force-cup). Remove the pop-up drain plug and strainer first, and plug the overflow hole by stuffing a wet rag into it, allowing you to create air pressure with the plunger.

Tools & Materials	Flashlight
Plunger	Rag
Channel-type pliers	Bucket
Small wire brush	Replacement gaskets
Screwdriver	Teflon tape

CLEARING A SINK TRAP

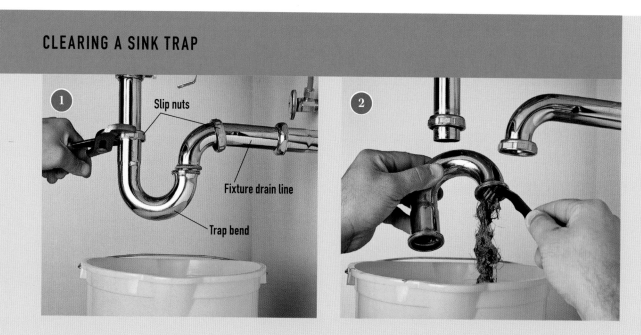

1   Place bucket under trap to catch water and debris. Loosen slip nuts on trap bend with channel-type pliers. Unscrew nuts by hand and slide away from connections. Pull off trap bend.

2   Dump out debris. Clean trap bend with a small wire brush. Inspect slip nut washers for wear and replace if necessary. Reinstall trap bend and tighten slip nuts.

CLEARING A KITCHEN SINK

Drainline from dishwasher

1   Plunging a kitchen sink is not difficult, but you need to create an uninterrupted pressure lock between the plunger and the clog. If you have a dishwasher, the drain tube needs to be clamped shut and sealed off at the disposer or drainline. The pads on the clamp should be large enough to flatten the tube across its full diameter (or you can clamp the tube ends between small boards).

2   If there is a second basin, have a helper hold a basket strainer plug in its drain or put a large pot or bucket full of water on top of it. Unfold the skirt within the plunger and place this in the drain of the sink you are plunging. There should be enough water in the sink to cover the plunger head. Plunge rhythmically for six repetitions with increasing vigor, pulling up hard on the last repetition. Repeat this sequence until the clog is removed. Flush out a cleared clog with plenty of hot water.

USING A HAND AUGER AT THE TRAP ARM

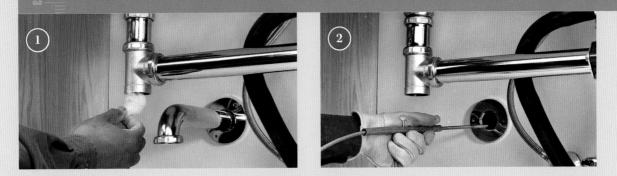

1   If plunging doesn't work, remove the trap and clean it out (see previous page). With the trap off, see if water flows freely from both sinks (if you have two). Sometimes clogs will lodge in the T-fitting or one of the waste pipes feeding it. These may be pulled out manually or cleared with a bottlebrush or wire. When reassembling the trap, apply Teflon tape clockwise to the male threads of metal waste pieces. Tighten with your channel-type pliers. Plastic pieces need no tape and should be hand tightened only.

2   If you suspect the clog is downstream of the trap, remove the trap arm from the fitting at the wall. Look in the fixture drain with a flashlight. If you see water, that means the fixture drain is plugged. Clear it with a hand-crank or drill-powered auger.

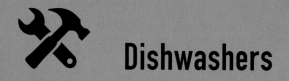

Dishwashers

Replacing an old, inefficient dishwasher is a straightforward project that usually takes just a few hours. The energy savings begin with the first load of dishes and continue with every load thereafter.

A DISHWASHER THAT'S PAST its prime may be inefficient in more ways than one. If it's an old model, it probably wasn't designed to be very efficient to begin with. But more significantly, if it no longer cleans effectively, you're probably spending a lot of time and hot water pre-rinsing the dishes. This alone can consume more energy and water than a complete wash cycle on a newer machine. So even if your old dishwasher still runs, replacing it with an efficient new model can be a good green upgrade.

In terms of sizing and utility hookups, dishwashers are generally quite standard. If your old machine is a built-in and your countertops and cabinets are standard sizes, most full-size dishwashers will fit right in. Of course, you should always measure the dimensions of the old unit before shopping for a new one to avoid an unpleasant surprise at installation time. Also be sure to review the manufacturer's instructions before starting any work.

REPLACING A DISHWASHER

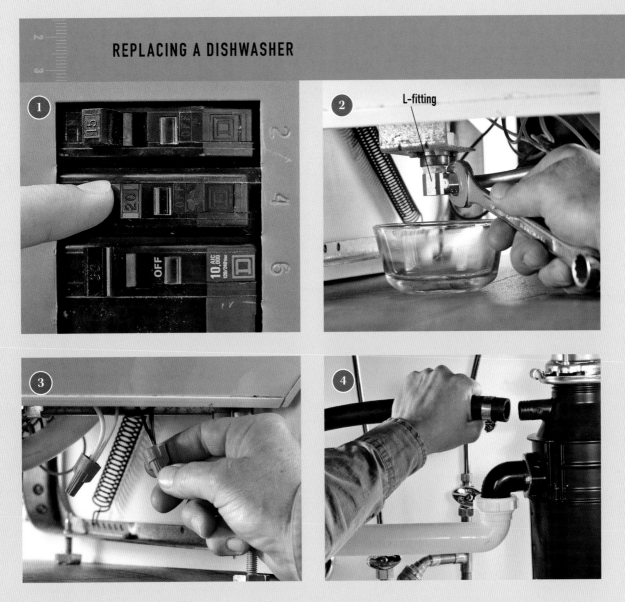

1   Start by shutting off the electrical power to the dishwasher circuit at the service panel. Also, turn off the water supply at the shutoff valve, usually located directly under the floor.

2   Disconnect old plumbing connections. First unscrew the front access panel. Once the access panel is removed, disconnect the water supply line from the L-fitting on the bottom of the unit. This is usually a brass compression fitting, so just turning the compression nut counterclockwise with an adjustable wrench should do the trick. Use a bowl to catch any water that might leak out when the nut is removed.

3   Disconnect old wiring connections. The dishwasher has an integral electrical box at the front of the unit where the power cable is attached to the dishwasher's fixture wires. Take off the box cover and remove the wire connectors that join the wires together.

4   Disconnect the discharge hose, which is usually connected to the dishwasher port on the side of the garbage disposer. To remove it, just loosen the screw on the hose clamp and pull it off. You may need to push this hose back through a hole in the cabinet wall and into the dishwasher compartment so it won't get caught when you pull the dishwasher out.

continued

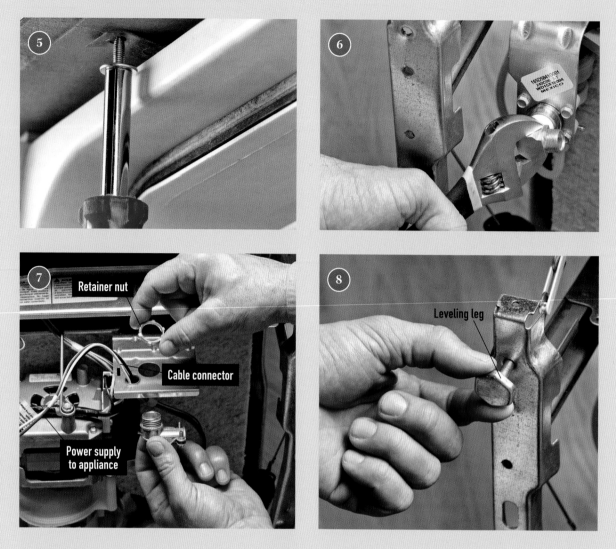

5   Detach the unit from the cabinets before you pull it out. Remove the screws that hold the brackets to the underside of the countertop. Then put a piece of cardboard or old carpet under the front legs to protect the floor from getting scratched, and pull the dishwasher out.

6   First, prepare the new dishwasher. Tip it on its back and attach the new L-fitting into the threaded port on the solenoid. Apply some Teflon tape or pipe sealant to the fitting threads before tightening it in place to prevent possible leaks.

7   Prepare for the wiring connections. Like the old dishwasher, the new one will have an integral electrical box for making the wiring connections. To gain access to the box, just remove the box cover. Then install a cable connector on the back of the box and bring the power cable from the service panel through this connector. Power should be shut off at the main service panel at all times.

8   Install a leveling leg at each of the four corners while the new dishwasher is still on its back. Just turn the legs into the threaded holes designed for them. Leave about ½" of each leg projecting from the bottom of the unit. These will have to be adjusted later to level the appliance. Tip the appliance up onto the feet and slide it into the opening. Check for level in both directions and adjust the feet as required.

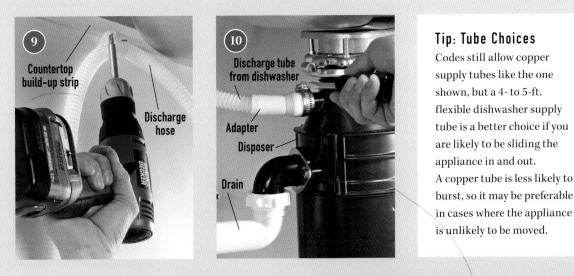

9 Countertop build-up strip / Discharge hose

10 Discharge tube from dishwasher / Adapter / Disposer / Drain

Tip: Tube Choices

Codes still allow copper supply tubes like the one shown, but a 4- to 5-ft. flexible dishwasher supply tube is a better choice if you are likely to be sliding the appliance in and out.

A copper tube is less likely to burst, so it may be preferable in cases where the appliance is unlikely to be moved.

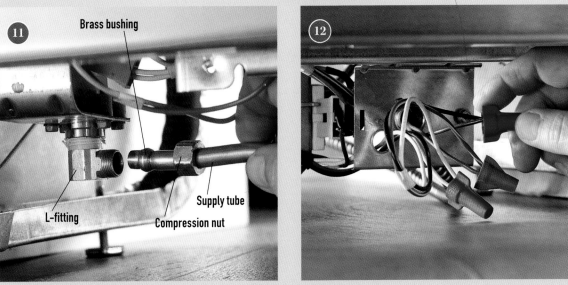

11 Brass bushing / L-fitting / Supply tube / Compression nut

12

9   Once the dishwasher is level, attach the brackets to the underside of the countertop to keep the appliance from moving. Then pull the discharge hose into the sink cabinet and install it so there's a loop that is attached with a bracket to the underside of the countertop. This loop prevents waste water from flowing from the disposer back into the dishwasher. Note: Some codes require that you install an air gap fitting for this purpose. Check with your local plumbing inspector.

10 Push an adapter over the disposer's discharge nipple and tighten it in place with a hose clamp. If you don't have a disposer, replace one of the drain tailpieces with a dishwasher tailpiece, and clamp the discharge tube to its fitting.

11 Adjust the L-fitting on the dishwasher's water inlet valve until it points directly toward the water supply tubing. Then lubricate the threads slightly with a drop of dishwashing liquid and tighten the tubing's compression nut onto the fitting. Keeping the brass bushing between the nut and the L-fitting. Use an adjustable wrench and turn the nut clockwise.

12 Complete the electrical connections by clamping the cable and joining the wires with wire nuts, following manufacturer's instructions. Replace the electrical cover, usually by hooking it onto a couple of prongs and driving a screw. Restore power and water, and test. Replace the toe-kick.

Food Disposers

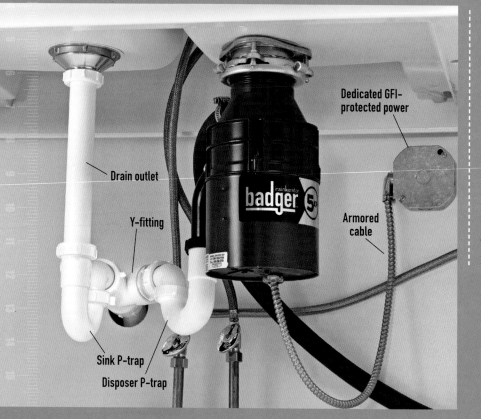

Drain outlet

Y-fitting

Sink P-trap

Disposer P-trap

Dedicated GFI-protected power

Armored cable

Tools & Materials

Screwdriver
Channel-type pliers
Spud wrench (optional)
Hammer
Hacksaw or tubing cutter
Kitchen drain supplies
Drain auger
Putty knife
Mineral spirits
Plumber's putty
Wire caps
Hose clamps
Threaded Y-fitting
Electrical tape

A properly functioning food disposer that's used correctly can help reduce clogs. Some plumbers use separate P-traps for the disposer and the drain outlet tube as shown here. Others contend that configuring the drain line with a single P-trap minimizes the chance that a trap will have its water seal broken by suction from the second trap.

FOOD DISPOSERS ARE standard equipment in the modern home, and most of us have come to depend on them to macerate our plate leavings and crumbs so they can exit the house along with waste water from the sink drain. If your existing disposer needs replacing, you'll find that the job is relatively simple, especially if you select a replacement appliance that is the same model as the old one. In that case, you can probably reuse the existing mounting assembly, drain sleeve, and drain plumbing.

Disposers are available with power ratings between 1/3 and 1 HP (horsepower). More powerful models

bog down less under load and the motors last longer because they don't have to work as hard. They are also costlier.

Choose a switch option that meets your family's safety needs. A "continuous feed" disposer may be controlled by a standard on-off switch on the wall. Another option is a disposer that stays on only when the switch is actively pressed. A "batch feed" disposer can turn on only when a lid is locked onto it, eliminating the possibility of harming fingers. Some models are controlled at the lid, without a wall switch. Continuous food disposers are the most common.

Shown cutaway

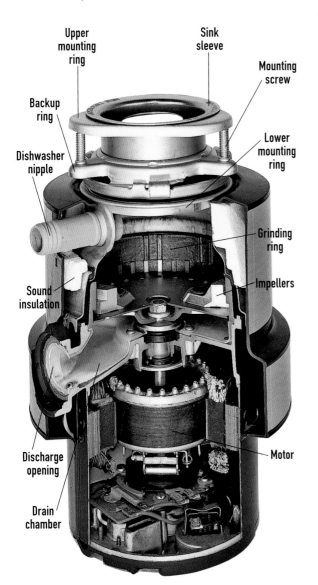

Upper mounting ring

Sink sleeve

Mounting screw

Backup ring

Lower mounting ring

Dishwasher nipple

Grinding ring

Sound insulation

Impellers

Discharge opening

Motor

Drain chamber

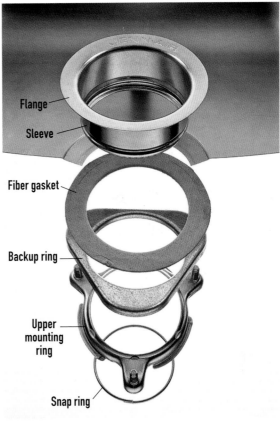

Flange

Sleeve

Fiber gasket

Backup ring

Upper mounting ring

Snap ring

The disposer is attached directly to the sink sleeve, which comes with the disposer and replaces the standard sink strainer. A snap ring fits into a groove around the sleeve of the strainer body to prevent the upper mounting ring and backup ring from sliding down while the upper mounting ring is tightened against the backup ring with mounting screws. A fiber gasket compresses when the mounting screws are tightened to create a better seal under the flange.

A food disposer grinds food waste so it can be flushed away through the sink drain system. A quality disposer has a ½ –horsepower, or larger, self-reversing motor. Other features to look for include foam sound insulation, a grinding ring, and overload protection that allows the motor to be reset if it overheats. Better food disposers have a 5-year manufacturer's warranty.

Kitchen and drain tees are required to have a baffle if the tee is connected to a dishwasher or disposer. The baffle is intended to prevent discharge from finding its way up the drain and into the sink.

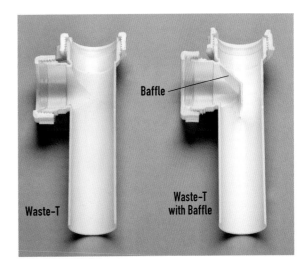

Baffle

Waste-T

Waste-T with Baffle

Mounting lug

Upper mounting ring

Lower mounting ring

Snap ring

Sink sleeve

1   Remove the old disposer if you have one. You'll need to disconnect the drain pipes and traps first. If your old disposer has a special wrench for the mounting lugs, use it to loosen the lugs. Otherwise, use a screwdriver. If you do not have a helper, place a solid object directly beneath the disposer to support it before you begin removal. Important: Shut off electrical power at the main service panel before you begin removal. Disconnect the wire leads, cap them, and stuff them into the electrical box.

2   Clear the drain lines all the way to the branch drain before you begin the new installation. Remove the trap and trap arm first.

3   Disassemble the mounting assembly and then separate the upper and lower mounting rings and the backup ring. Also remove the snap ring from the sink sleeve. See photo, previous page.

4   Press the flange of the sink sleeve for your new disposer into a thin coil of plumber's putty that you have laid around the perimeter of the drain opening. The sleeve should be well-seated in the coil.

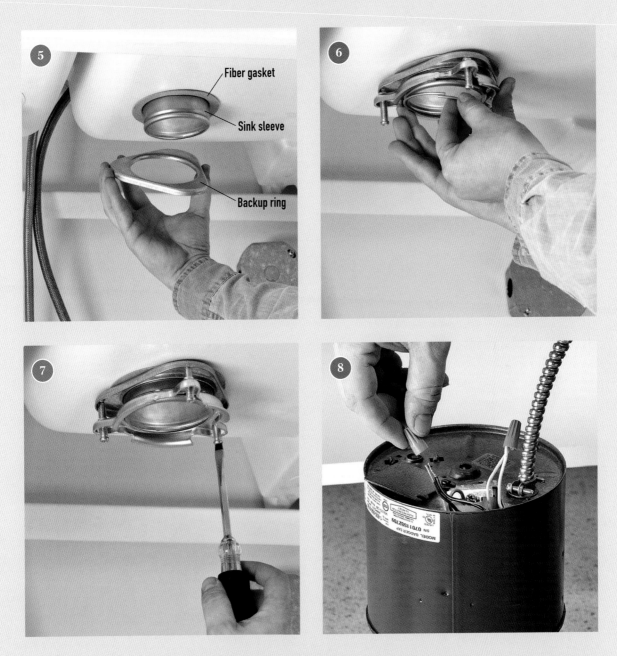

5   Slip the fiber gasket and then the backup ring onto the sink sleeve, working from inside the sink base cabinet. Make sure the backup ring is oriented the same way it was before you disassembled the mounting assembly.

6   Insert the upper mounting ring onto the sleeve with the slotted ends of the screws facing away from the backup ring so you can access them. Then, holding all three parts at the top of the sleeve, slide the snap ring onto the sleeve until it snaps into the groove.

7   Tighten the three mounting screws on the upper mounting ring until the tips press firmly against the backup ring. It is the tension created by these screws that keeps the disposer steady and minimizes vibrating.

8   Make electrical connections before you mount the disposer unit on the mounting assembly. Shut off the power at the service panel if you have turned it back on. Remove the access plate from the disposer. Attach the white and black feeder wires from the electrical box to the white and black wires (respectively) inside the disposer. Twist a small wire cap onto each connection and wrap it with electrical tape for good measure. Also attach the green ground wire from the box to the grounding terminal on your disposer.

continued

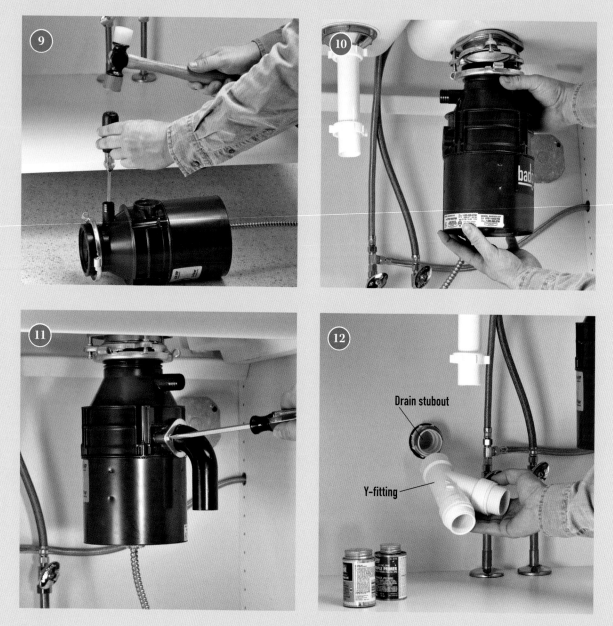

Drain stubout

Y-fitting

9   Knock out the plug in the disposer port if you will be connecting your dishwasher to the disposer. If you have no dishwasher, leave the plug in. Insert a large flathead screwdriver into the port opening and rap it with a mallet. Retrieve the knock-out plug from inside the disposer canister.

10 Hang the disposer from the mounting ring attached to the sink sleeve. To hang it, simply lift it up and position the unit so the three mounting ears are underneath the three mounting screws and then spin the unit so all three ears fit into the mounting assembly. Wait until after the plumbing hookups have been made to lock the unit in place.

11 Attach the discharge tube to the disposer according to the manufacturer's instructions. It is important to get a very good seal here, or the disposer will leak. Go ahead and spin the disposer if it helps you access the discharge port.

12 Attach a Y-fitting at the drain stubout. The Y-fitting should be sized to accept a drain line from the disposer and another from the sink. Adjust the sink drain plumbing as needed to get from the sink P-trap to one opening of the Y.

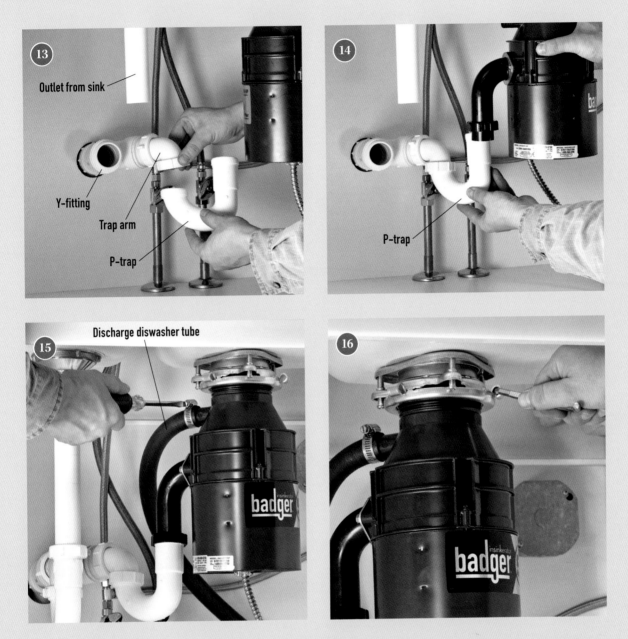

Outlet from sink

Y-fitting

Trap arm

P-trap

P-trap

Discharge diswasher tube

13 Install a trap arm for the disposer in the open port of the Y-fitting at the wall stubout. Then, attach a P-trap or a combination of a tube extension and a P-trap so the trap will align with the bottom of the disposer discharge tube.

14 Spin the disposer so the end of the discharge tube is lined up over the open end of the P-trap and confirm that they will fit together correctly. If the discharge tube extends down too far, mark a line on it at the top of the P-trap and cut at the line with a hacksaw. If the tube is too short, attach an extension with a slip joint. You may need to further shorten the discharge tube first to create enough room for the slip joint on the extension. Slide a slip nut and beveled compression washer onto the discharge tube and attach the tube to the P-trap.

15 Connect the dishwasher discharge tube to the inlet port located at the top of the disposer unit. This may require a dishwasher hookup kit. Typically, a hose clamp is used to secure the connection.

16 Lock the disposer into position on the mounting ring assembly once you have tested to make sure it is functioning correctly and without leaks. Lock it by turning one of the mounting lugs until it makes contact with the locking notch.

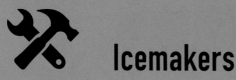

Icemakers

Tools & Materials

Screwdrivers
Nut drivers
Needle-nose pliers
Duct or masking tape
Channel-type pliers
Electric drill and assorted bits
Icemaker kit (or icemaker tubing
   with ferrules and nuts)
Open-end or adjustable wrench
T-fitting (for supply tube)
Putty knife
Long ½"-inch drill bit
Electrician's tape

A built-in icemaker is easy to install as a retrofit appliance in most modern refrigerators. If you want to have an endless supply of ice for home use, you'll wonder how you ever got along without one.

MANY REFRIGERATORS COME with icemakers as standard equipment, and practically every model features them as an option (a refrigerator with an icemaker usually costs about $100 more). It is also possible to purchase an icemaker as a retrofit feature for your old fridge.

Hooking up an existing icemaker to a cold-water supply involves drilling holes and connecting to a cold-water pipe. Most often, a pipe can be found in the basement below the kitchen, perhaps under the kitchen sink. To make the connection, some local codes allow the installation of a saddle Tee valve, but many do not, and a compression Tee valve is not difficult to install, as we show. In many kitchens the flexible line running from the valve to the fridge is

copper, but plastic icemaker tubing is easier to install and less likely to kink or crack. To be sure everything fits, you can buy a connection kit from the refrigerator manufacturer.

Most icemakers either come preinstalled or are purchased as an accessory when you buy your new refrigerator. But if you have an older refrigerator with no icemaker and you'd like it to have one, all is not lost. Inspect the back of the unit, behind the freezer compartment. If your refrigerator has the required plumbing to support an icemaker, you will see a port or a port that is covered with backing. In that case, all you need to do is take the make and model information to an appliance parts dealer and they can sell you an aftermarket icemaker. Plan to spend $100 to $200.

How Icemakers Work

An icemaker receives its supply of water for making cubes through a ¼" copper supply line that runs from the icemaker to a water pipe. The supply line runs through a valve in the refrigerator and is controlled by a solenoid that monitors the water supply and sends the water into the icemaker itself, where it is turned into ice cubes. The cubes drop down into a bin, and as the ice level rises, they also raise a bail wire that's connected to a shutoff. When the bin is full, the bail wire will be high enough to trigger a mechanism that shuts off the water supply.

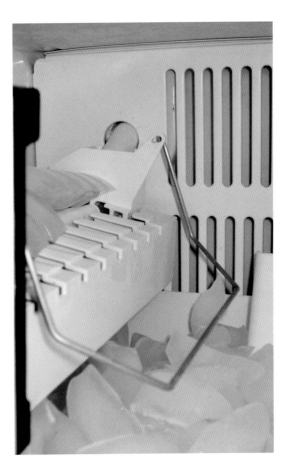

Aftermarket automatic icemakers are simple to install as long as your refrigerator is icemaker ready. Make sure to buy the correct model for your appliance and do careful installation work— icemaker water supply lines are very common sources for leaks.

CONNECTING AN ICEMAKER

1   Locate a nearby cold-water pipe, usually in the basement or crawl space below the kitchen. Behind the refrigerator and near the wall, use a long ½" bit to drill a hole through the floor. Do not pull the bit out.

continued

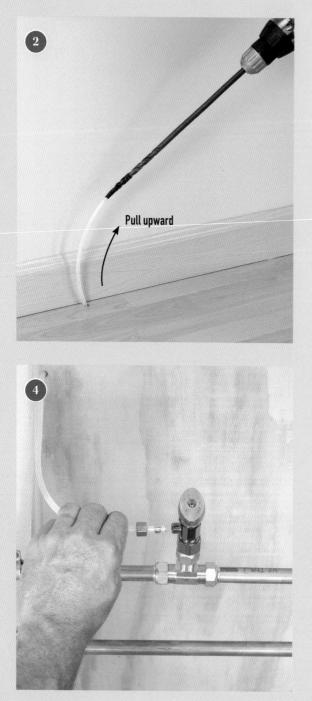

Pull upward

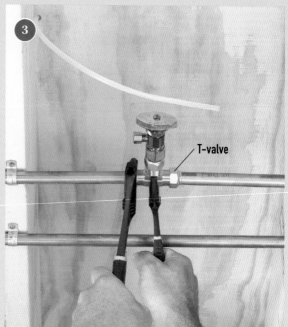

T-valve

2   From below, fasten plastic icemaker tubing to the end of the drill bit by wrapping firmly with electrician's tape. From above, carefully pull the bit up, to thread the tubing up into the kitchen.

3   Shut off the water and open nearby faucets to drain the line. Cut into a cold-water pipe and install a compression Tee valve. Tighten all the nuts, close the valve and nearby faucets, and restore water to test for leaks.

4   Connect the tubing. Arrange the tubing behind the fridge so you have about 6 ft. of slack, making it easy to pull the fridge out for cleaning. Cut the tubing with a knife. Slide on a nut and a ferrule. Insert the tubing into the valve, slide the ferrule tight against the valve, and tighten the nut. To finish the installation, connect the tubing to the refrigerator using a nut and ferrule. Keep the tubing neatly coiled and kink-free for future maintenance.

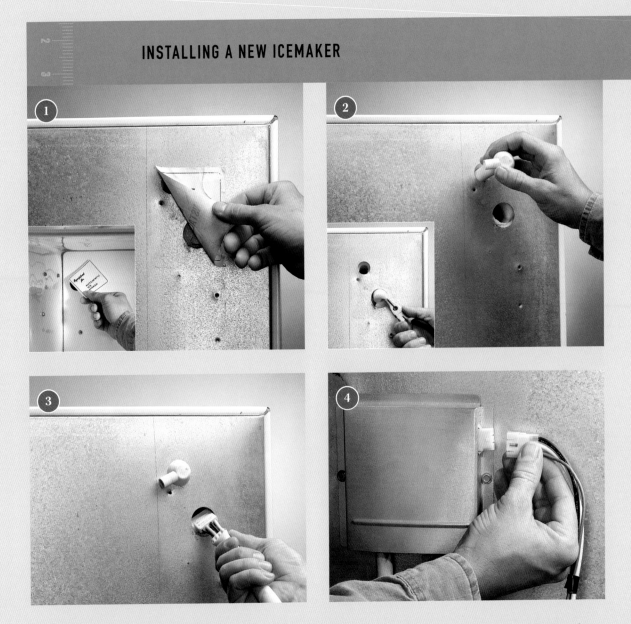

1   Remove all the contents from the refrigerator and freezer. Unplug the unit and pull it out from the wall. Open the freezer door and remove the icemaker cover plate (inset). On the back of the refrigerator, remove the backing or unscrew the icemaker access panel.

2   Install the tube assembly. Remove two insulation plugs to expose two openings, one for the water line and the other for a wiring harness. Install the water tube assembly (part of the icemaker kit) in its access hole; it has a plastic elbow attached to the plastic tube that reaches into the freezer compartment.

3   Hook up the harness. Icemaker kits usually come with a wiring harness that joins the icemaker motor inside the freezer box to the power supply wires. Push this harness through its access hole and into the freezer compartment. Then seal the hole with the plastic grommet that comes with the harness.

4   Join the end of the icemaker wiring harness to the power connector that was preinstalled on the back of the refrigerator. This connection should lay flat against the back. If it doesn't, just tape it down with some duct tape or masking tape.

continued

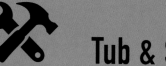

Tub & Shower Faucets

Tools & Materials

Screwdrivers

Nut drivers

Needle-nose pliers

Duct or masking tape

Channel-type pliers

Electric drill and assorted bits

Icemaker kit (or icemaker tubing
     with ferrules and nuts)

Open-end or adjustable wrench

T-fitting (for supply tube)

Putty knife

Long ½-inch drill bit

Electrician's tape

Tub/shower plumbing is notorious for developing drips from the tub spout and the showerhead. In most cases, the leak can be traced to the valves controlled by the faucet handles.

TUB AND SHOWER FAUCETS have the same basic designs as sink faucets, and the techniques for repairing leaks are the same as described in the faucet repair section of this book (pages xx to xx). To identify your faucet design, you may have to take off the handle and disassemble the faucet.

When a tub and shower are combined, the showerhead and the tub spout share the same hot and cold water supply lines and handles. Combination faucets are available as three-handle, two-handle, or single-handle types. The number of handles gives clues as to the design of the faucets and the kinds of repairs that may be necessary.

With combination faucets, a diverter valve or gate diverter is used to direct water flow to the tub spout or the showerhead. On three-handle faucet types, the middle handle controls a diverter valve. If water does not shift easily from tub to showerhead, or if water continues to run out the spout when the

shower is on, the diverter valve probably needs to be cleaned and repaired.

Two-handle and single-handle types use a gate diverter that is operated by a pull lever or knob on the tub spout. Although gate diverters rarely need repair, the lever occasionally may break, come loose, or refuse to stay in the up position. To repair a gate diverter set in a tub spout, replace the entire spout.

Tub and shower faucets and diverter valves may be set inside wall cavities. Removing them may require a deep-set ratchet wrench.

If spray from the showerhead is uneven, clean the spray holes. If the showerhead does not stay in an upright position, remove the head and replace the O-ring.

To add a shower to an existing tub, install a flexible shower adapter. Several manufacturers make complete conversion kits that allow a shower to be installed in less than one hour.

Tub & Shower Combination Faucets

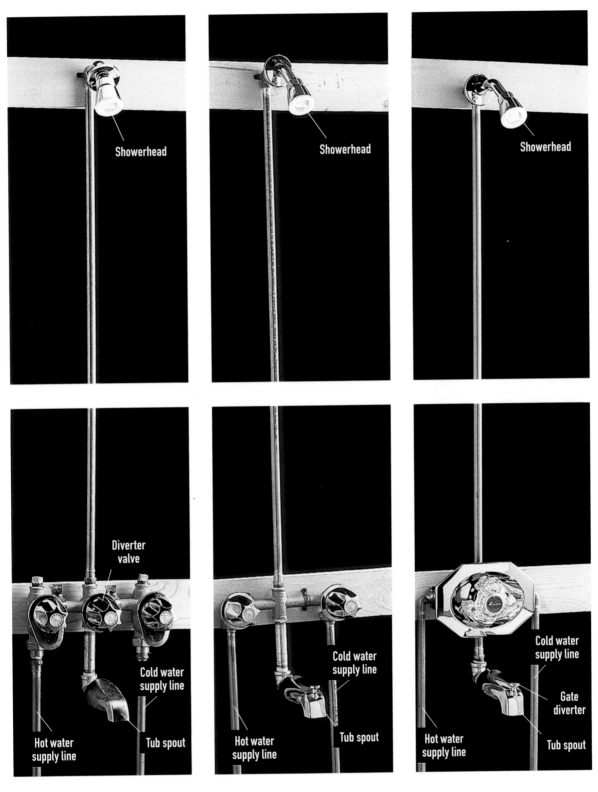

Three-handle faucet has valves that are either compression or cartridge design.

Two-handle faucet has valves that are either compression or cartridge design.

Single-handle faucet has valves that are cartridge, ball-type, or disc design.

Repair Three-handle Tub & Shower Faucets

A three-handle faucet type has two handles to control hot and cold water, and a third handle to control the diverter valve and direct water to either a tub spout or a shower head. The separate hot and cold handles indicate cartridge or compression faucet designs.

If a diverter valve sticks, if water flow is weak, or if water runs out of the tub spout when the flow is directed to the showerhead, the diverter needs to be repaired or replaced. Most diverter valves are similar to either compression or cartridge faucet valves. Compression-type diverters can be repaired, but cartridge types should be replaced.

Tools & Materials

Screwdriver	Replacement diverter
Adjustable wrench or	cartridge or
channel-type pliers	universal washer kit
Deep-set ratchet wrench	Faucet grease
Small wire brush	Vinegar

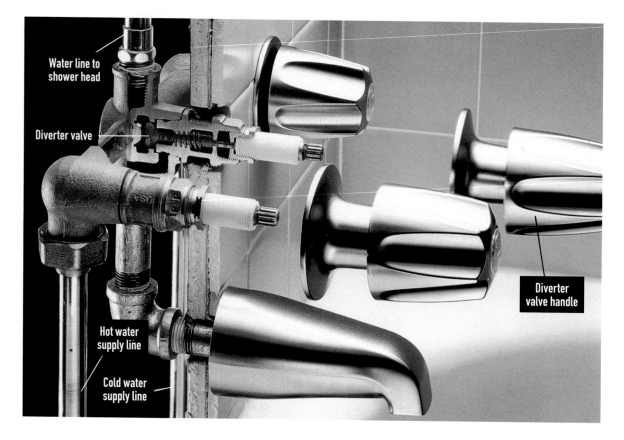

Water line to shower head

Diverter valve

Hot water supply line

Cold water supply line

Diverter valve handle

A three-handle tub/shower faucet has individual controls for hot and cold water plus a third handle that operates the diverter valve.

1   Remember to turn off the water before beginning work. Remove the diverter valve handle with a screwdriver. Unscrew or pry off the escutcheon.

2   Remove bonnet nut with an adjustable wrench or channel-type pliers.

3   Unscrew the stem assembly, using a deep-set ratchet wrench. If necessary, chip away any mortar surrounding the bonnet nut.

4   Remove the brass stem screw. Replace the stem washer with an exact duplicate. If the stem screw is worn, replace it.

5   Unscrew the threaded spindle from the retaining nut.

6   Clean sediment and lime buildup from the nut using a small wire brush dipped in vinegar. Coat all parts with faucet grease and reassemble the diverter valve.

Repair Two-handle Tub & Shower Faucets

Two-handle tub and shower faucets are either cartridge or compression design. Because the valves of two-handle tub and shower faucets may be set inside the wall cavity, a deep-set socket wrench may be required to remove the valve stem.

Two-handle tub and shower designs have a gate diverter. A gate diverter is a simple mechanism located in the tub spout. A gate diverter closes the supply of water to the tub spout and redirects the flow to the shower head. Gate diverters seldom need repair. Occasionally, the lever may break, come loose, or refuse to stay in the up position.

If the diverter fails to work properly, replace the tub spout. Tub spouts are inexpensive and easy to replace.

Tools & Materials
Screwdriver
Allen wrench
Pipe wrench
Channel-type pliers
Small cold chisel
Ball-peen hammer
Deep-set ratchet wrench
Masking tape or cloth
Pipe joint compound
Replacement faucet
    parts, as needed

A two-handle tub/shower faucet can operate with compression valves, but more often these days they contain cartridges that can be replaced. Unlike a three-handled model, the diverter is a simple gate valve that is operated by a lever.

Tips on Replacing a Tub Spout

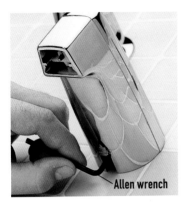

Allen wrench

Spout nipple

Check underneath tub spout for a small access slot. The slot indicates the spout is held in place with an Allen screw. Remove the screw using an Allen wrench. Spout will slide off.

Unscrew faucet spout. Use a pipe wrench, or insert a large screwdriver or hammer handle into the spout opening and turn spout counterclockwise.

Spread pipe joint compound on threads of spout nipple before replacing spout. If you have a copper pipe or a short pipe, buy a spout retrofit kit, which can attach a spout to most any pipe.

REMOVING A DEEP-SET FAUCET VALVE

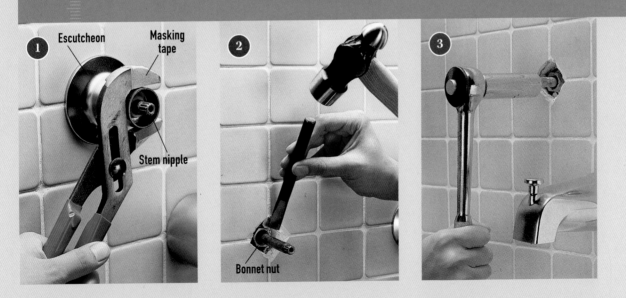

1 Remember to turn off the water before beginning any work. Remove the handle and unscrew the escutcheon with channel-type pliers. Pad the jaws of the pliers with masking tape to prevent scratching the escutcheon.

2 Chip away any mortar surrounding the bonnet nut using a ball-peen hammer and a small cold chisel.

3 Unscrew the bonnet nut with a deep-set ratchet wrench. Remove the bonnet nut and stem from the faucet body.

Repair Single-handle Tub & Shower Faucets

A single-handle tub and shower faucet has one valve that controls both water flow and temperature. Single-handle faucets may be ball-type, cartridge, or disc designs.

If a single-handle control valve leaks or does not function properly, disassemble the faucet, clean the valve, and replace any worn parts. Use the repair techniques described on page 80 for ball-type, or page 81 for ceramic disc. Repairing a single-handle cartridge faucet is shown on the opposite page.

Direction of the water flow to either the tub spout or the showerhead is controlled by a gate diverter. Gate diverters seldom need repair. Occasionally, the lever may break, come loose, or refuse to stay in

the up position. Remember to turn off the water before beginning any work; the shower faucet shown here has built-in shutoff valves, but many other valves do not. Open an access panel in an adjoining room or closet, behind the valve, and look for two shutoffs. If you can't find them there, you may have to shut off intermediate valves or the main shutoff valve.

Tools & Materials
Screwdriver
Adjustable wrench

Channel-type pliers
Replacement faucet
parts, as needed

Water line to
shower head

Built-in shutoff valves

Escutcheon

Control valve

Cold water
supply line

Gate diverter

Hot water
supply line

A single-handle tub/shower faucet is the simplest type to operate and to maintain. The handle controls the mixing ratio of both hot and cold water, and the diverter is a simple gate valve.

REPAIRING A SINGLE-HANDLE CARTRIDGE TUB & SHOWER FAUCET

1   Use a screwdriver to remove the handle and escutcheon.

2   Turn off water supply at the built-in shutoff valves or the main shutoff valve.

3   Unscrew and remove the retaining ring or bonnet nut using adjustable wrench.

4   Remove the cartridge assembly by grasping the end of the valve with channel-type pliers and pulling gently.

5   Flush the valve body with clean water to remove sediment. Replace any worn O-rings. Reinstall the cartridge and test the valve. If the faucet fails to work properly, replace the cartridge.

Single-Handle Tub & Shower Faucet with Scald Control

In many plumbing systems, if someone flushes a nearby toilet or turns on the cold water of a nearby faucet while someone else is taking a shower, the shower water temperature can suddenly rise precipitously. This is not only uncomfortable; it can actually scald you. For that reason, many one-handle shower valves have a device, called a "balancing valve" or an "anti-scald valve," that keeps the water from getting too hot.

The temperature of your shower may drastically rise to dangerous scalding levels if a nearby toilet is flushed. A shower fixture equipped with an anti-scald valve prevents this sometimes dangerous situation.

ADJUSTING THE SHOWER'S TEMPERATURE

1   To reduce or raise the maximum temperature, remove the handle and escutcheon. Some models have an adjustment screw, others have a handle that can be turned by hand.

2   To remove a balancing valve, you may need to buy a removal tool made for your faucet. Before replacing, slowly turn on water to flush out any debris; use a towel or bucket to keep water from entering inside the wall.

Fix & Replace Showerheads

If spray from the showerhead is uneven, clean the spray holes. The outlet or inlet holes of the showerhead may get clogged with mineral deposits. Showerheads pivot into different positions. If a showerhead does not stay in position, or if it leaks, replace the O-ring that seals against the swivel ball.

A tub can be equipped with a shower by installing a flexible shower adapter kit. Complete kits are available at hardware stores and home centers.

Tools & Materials

Adjustable wrench or channel-type pliers
Pipe wrench
Drill
Glass and tile bit
Mallet
Screwdriver
Masking tape
Thin wire (paper clip)
Faucet grease
Rag
Replacement O-rings
Masonry anchors
Flexible shower adapter kit (optional)

A typical showerhead can be disassembled easily for cleaning and repair. Some showerheads include a spray adjustment cam lever that is used to change the force of the spray.

CLEANING & REPAIRING A SHOWERHEAD

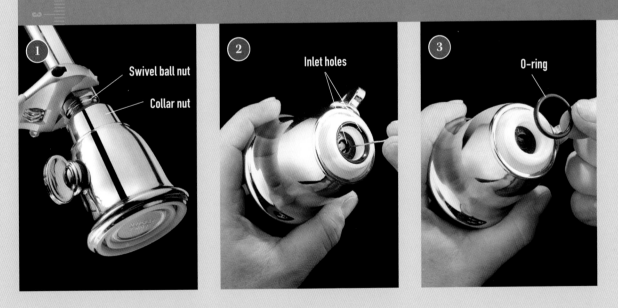

1   Unscrew the swivel ball nut, using an adjustable wrench or channel-type pliers. Wrap jaws of the tool with masking tape to prevent marring the finish. Unscrew the collar nut from the showerhead.

2   Clean outlet and inlet holes of showerhead with a thin wire. Flush the head with clean water.

3   Replace the O-ring, if necessary. Lubricate the O-ring with faucet grease before installing.

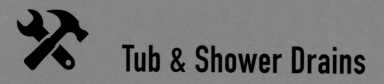

Tub & Shower Drains

Tools & Materials
Phillips screwdriver
Plunger
Scrub brush
White vinegar
Toothbrush
Needlenose pliers
Dishwashing brush
Faucet grease

As with bathroom sinks, tub and shower drain pipes may become clogged with soap and hair. The drain stopping mechanisms can also require cleaning and adjustment.

TUB OR SHOWER NOT DRAINING? First, make sure it's only the tub or shower. If your sink is plugged, too, it may be a coincidence or it may be that a common branch line is plugged. A sure sign of this is when water drains from the sink into the tub. This could require the help of a drain cleaning service, or a drum trap that services both the sink and tub needs cleaning.

If the toilet also can't flush (or worse, water comes into the tub when you flush the toilet), then the common drain to all your bathroom fixtures is plugged. Call a drain cleaning service. If you suspect the problem is only with your tub or shower, then read on. We'll show you how to clear drainlines and clean and adjust two types of tub stopper mechanisms. Adjusting the mechanism can also help with the opposite problem: a tub that drains when you're trying to take a bath.

Maintenance Tip
Like bathroom sinks, tubs and showers face an ongoing onslaught from soap and hair. When paired, this pesky combination is a sure-fire source of clogs. The soap scum coagulates as it is washed down the drain and binds the hair together in a mass that grows larger with every shower or bath. To nip these clogs in the bud, simply pour boiling hot clean water down the drain from time to time to melt the soapy mass and wash the binder away.

On shower drains, feed the head of a hand-crank or drill-powered auger in through the drain opening after removing the strainer. Crank the handle of the auger to extend the cable and the auger head down into the trap and, if the clog is farther downline, toward the branch drain. When clearing any drain, it is always better to retrieve the clog than to push it farther downline.

Sloped floor

Floor

Drain opening

Trap arm

Trap

Branch drain line

On combination tub/showers, it's generally easiest to insert the auger through the overflow opening after removing the coverplate and lifting out the drain linkage. Crank the handle of the auger to extend the cable and the auger head down into the trap and, if the clog is farther downline, toward the branch drain. When clearing any drain, it is always better to retrieve the clog than to push it farther downline.

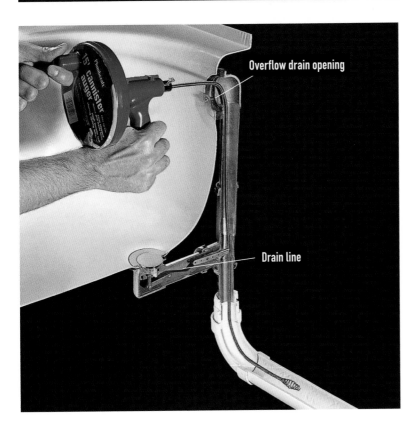

Overflow drain opening

Drain line

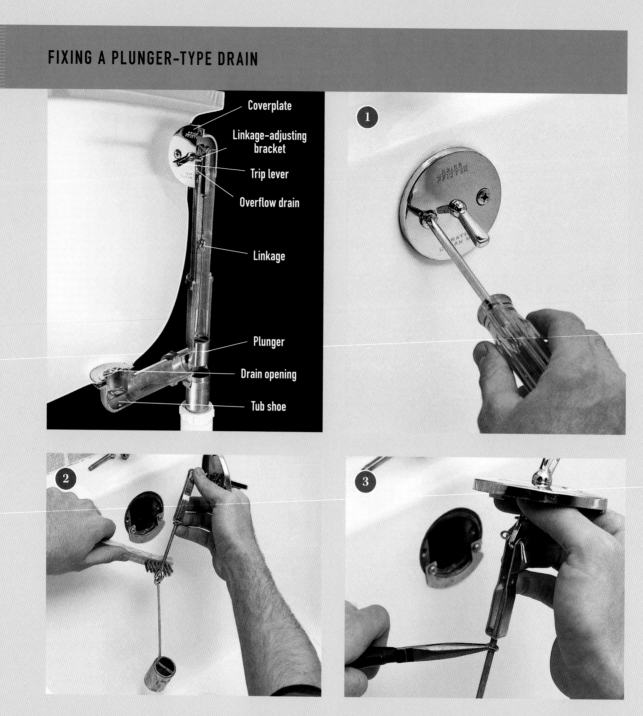

Coverplate

Linkage-adjusting bracket

Trip lever

Overflow drain

Linkage

Plunger

Drain opening

Tub shoe

1   A plunger-type tub drain has a simple grate over the drain opening and a behind-the-scenes plunger stopper. Remove the screws on the overflow coverplate with a slotted or Phillips screwdriver. Pull the coverplate, linkage, and plunger from the overflow opening.

2   Clean hair and soap off the plunger with a scrub brush. Mineral buildup is best tackled with white vinegar and a toothbrush or a small wire brush.

3   Adjust the plunger. If your tub isn't holding water with the plunger down, it's possible the plunger is hanging too high to fully block water from the tub shoe. Loosen the locknut with needlenose pliers, then screw the rod down about ⅛". Tighten the locknut down. If your tub drains poorly, the plunger may be set too low. Loosen the locknut and screw the rod in ⅛" before retightening the locknut.

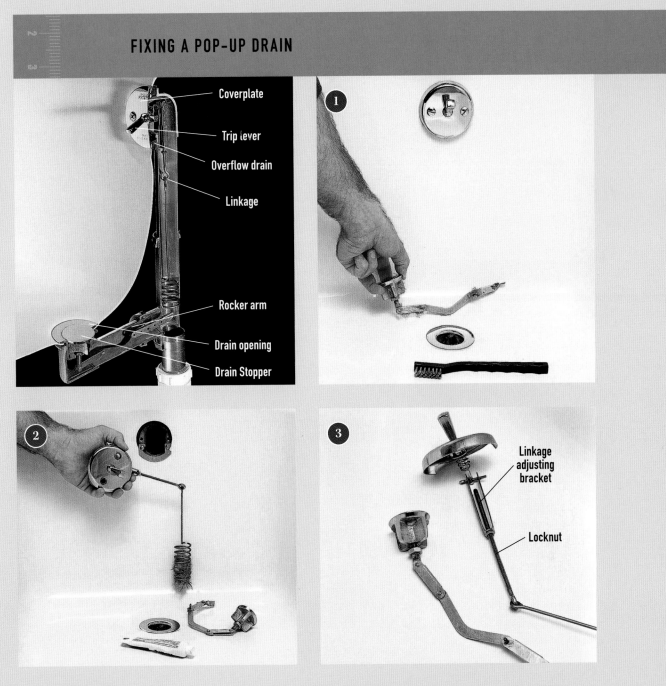

Coverplate

Trip lever

Overflow drain

Linkage

Rocker arm

Drain opening

Drain Stopper

Linkage adjusting bracket

Locknut

1   Raise the trip lever to the open position. Pull the stopper and rocker arm assembly from the drain. Clean off soap and hair with a dishwashing brush in a basin of hot water. Clean off mineral deposits with a toothbrush or small wire brush and white vinegar.

2   Remove the screws from the cover plate. Pull the trip lever and the linkage from the overflow opening. Clean off soap and hair with a brush in a basin of hot water. Remove mineral buildup with white vinegar and a wire brush. Lubricate moving parts of the linkage and rocker arm mechanism with faucet grease.

3   Adjust the pop-up stopper mechanism by first loosening the locknut on the lift rod. If the stopper doesn't close all the way, shorten the linkage by screwing the rod ⅛" farther into the linkage-adjusting bracket. If the stopper doesn't open wide enough, extend the linkage by unscrewing the rod ⅛". Tighten the locknut before replacing the mechanism and testing your adjustment.

Glossary

Access panel — Opening in a wall or ceiling that provides access to the plumbing system

Air admittance valve — A valve that allows air into a drain line in order to facilitate proper draining. Often used where traditional vent pipe would be difficult to install.

Appliance — Powered device that uses water, such as a water heater, dishwasher, washing machine, whirlpool, or water softener

Auger — Flexible tool used for clearing obstructions in drain lines

Ballcock — Valve that controls the water supply entering a toilet tank

Blow bag — Expanding rubber device that attaches to a garden hose; used for clearing floor drains

Branch drain line — Pipe that connects additional lines to a drain system

Branch line — Pipe that connects additional lines to a water supply system

Cleanout — Cover in a waste pipe or trap that provides access for cleaning

Closet auger — Flexible rod used to clear obstructions in toilets

Closet bend — Curved fitting that fits between a closet flange and a toilet drain

Closet flange — Ring at the opening of a toilet drain, used as the base for a toilet

Coupling — Fitting that connects two pieces of pipe

DWV — Drain, waste, and vent; the system for removing water from a house

DWV stack — Pipe that connects house drain system to a sewer line at the bottom and vents air to outside of house at the top

Elbow — Angled fitting that changes the direction of a pipe

Fixture — Device that uses water, such as a sink, tub, shower, sillcock, or toilet

Flapper (tank ball) — Rubber seal that controls the flow of water from a toilet tank to a toilet bowl

Flux (soldering paste) — Paste applied to metal joints before soldering to increase joint strength

Hand auger (snake) — Hand tool with flexible shaft, used for clearing clogs in drain lines

Hose bib — Any faucet spout that is threaded to accept a hose

I.D. — Inside diameter; plumbing pipes are classified by I.D.

Loop vent — A special type of vent configuration used in kitchen sink island installations

Main shutoff valve — Valve that controls water supply to an entire house; usually next to the water meter

Motorized auger — Power tool with flexible shaft, used for clearing tree roots from sewer lines

Nipple — Pipe with threaded ends

O.D. — Outside diameter

Plumber's putty — A soft material used for sealing joints between fixtures and supply or drain parts

Reducer — A fitting that connects pipes of different sizes

Riser — Assembly of water supply fittings and pipes that distributes water upward

Run — Assembly of pipes that extends from water supply to fixture, or from drain to stack

Saddle valve — Fitting clamped to copper supply pipe, with hollow spike that punctures the pipe to divert water to another device, usually a dishwasher or refrigerator icemaker

Sanitary fitting — Fitting that joins DWV pipes; allows solid material to pass through without clogging

Shutoff valve — Valve that controls the water supply for one fixture or appliance

Sillcock — Compression faucet used on the outside of a house

Soil stack — Main vertical drain line, which carries waste from all branch drains to a sewer line

Solder — Metal alloy used for permanently joining metal (usually copper) pipes

T-fitting — Fitting shaped like the letter T used for creating or joining branch lines

Trap — Curved section of drain, filled with standing water, that prevents sewer gases from entering a house

Union — Fitting that joins two sections of pipe but can be disconnected without cutting

Vacuum breaker — Attachment for outdoor and below-ground fixtures that prevents waste water from entering supply lines if water supply pressure drops

Wet vent — Pipe that serves as a drain for one fixture and as a vent for another

Y-fitting — Fitting shaped like the letter Y used for creating or joining branch lines

Metric Conversions

Metric Equivalent

Inches (in.)	1/64	1/32	1/25	1/16	1/8	1/4	3/8	2/5	1/2	5/8	3/4	7/8	1	2	3	4	5	6	7	8	9	10	11	12	36	39.4
Feet (ft.)																								1	3	3 1/12
Yards (yd.)																									1	1 1/12
Millimeters (mm)	0.40	0.79	1	1.59	3.18	6.35	9.53	10	12.7	15.9	19.1	22.2	25.4	50.8	76.2	101.6	127	152	178	203	229	254	279	305	914	1,000
Centimeters (cm)							0.95	1	1.27	1.59	1.91	2.22	2.54	5.08	7.62	10.16	12.7	15.2	17.8	20.3	22.9	25.4	27.9	30.5	91.4	100
Meters (m)																								.30	.91	1.00

Converting Measurements

To Convert:	To:	Multiply by:		To Convert:	To:	Multiply by:
Inches	Millimeters	25.4		Millimeters	Inches	0.039
Inches	Centimeters	2.54		Centimeters	Inches	0.394
Feet	Meters	0.305		Meters	Feet	3.28
Yards	Meters	0.914		Meters	Yards	1.09
Miles	Kilometers	1.609		Kilometers	Miles	0.621
Square inches	Square centimeters	6.45		Square centimeters	Square inches	0.155
Square feet	Square meters	0.093		Square meters	Square feet	10.8
Square yards	Square meters	0.836		Square meters	Square yards	1.2
Cubic inches	Cubic centimeters	16.4		Cubic centimeters	Cubic inches	0.061
Cubic feet	Cubic meters	0.0283		Cubic meters	Cubic feet	35.3
Cubic yards	Cubic meters	0.765		Cubic meters	Cubic yards	1.31
Pints (U.S.)	Liters	0.473 (Imp. 0.568)		Liters	Pints (U.S.)	2.114 (Imp. 1.76)
Quarts (U.S.)	Liters	0.946 (Imp. 1.136)		Liters	Quarts (U.S.)	1.057 (Imp. 0.88)
Gallons (U.S.)	Liters	3.785 (Imp. 4.546)		Liters	Gallons (U.S.)	0.264 (Imp. 0.22)
Ounces	Grams	28.4		Grams	Ounces	0.035
Pounds	Kilograms	0.454		Kilograms	Pounds	2.2
Tons	Metric tons	0.907		Metric tons	Tons	1.1

Converting Temperatures

Convert degrees Fahrenheit (F) to degrees Celsius (C) by following this simple formula: Subtract 32 from the Fahrenheit temperature reading. Then mulitply that number by $5/9$. For example, 77°F - 32 = 45. $45 \times 5/9 = 25$°C.

To convert degrees Celsius to degrees Fahrenheit, multiply the Celsius temperature reading by $9/5$, then add 32. For example, 25°C $\times 9/5 = 45$. 45 + 32 = 77°F.

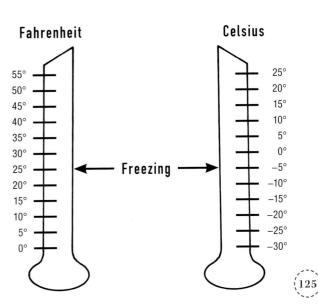

Index

First published in 2013 by Cool Springs Press, an imprint of the Quayside Publishing Group, 400 First Avenue North, Suite 400, Minneapolis, MN 55401

Cool Springs Press titles are also available at discounts in bulk quantity for industrial or sales-promotional use. For details write to Special Sales Manager at Cool Springs Press, 400 First Avenue North, Suite 400, Minneapolis, MN 55401 USA. To find out more about our books, visit us online at www.coolspringspress.com.

Library of Congress Cataloging-in-Publication Data

Homeskills. Plumbing : install & repair your own toilets, faucets, sinks, tubs, showers, drains.
    pages cm
  ISBN 978-1-59186-583-4 (softcover)
 1. Plumbing--Amateurs' manuals. I. Cool Springs Press. II. Title: Plumbing. III. Title: Home skills.

  TH6124.H613185 2013
  696'.1--dc23

                    2013005618

Design Manager: Cindy Samargia Laun
Design and layout: Kim Winscher
Cover and series design: Carol Holtz

Printed in China
10 9 8 7 6 5 4 3 2 1

NOTICE TO READERS

For safety, use caution, care, and good judgment when following the procedures described in this book. The publisher cannot assume responsibility for any damage to property or injury to persons as a result of misuse of the information provided.

The techniques shown in this book are general techniques for various applications. In some instances, additional techniques not shown in this book may be required. Always follow manufacturers' instructions included with products, since deviating from the directions may void warranties. The projects in this book vary widely as to skill levels required: some may not be appropriate for all do-it-yourselfers, and some may require professional help.

Consult your local building department for information on building permits, codes, and other laws as they apply to your project.